To Feed a Nation

Other Young People's Books by J. J. McCoy

SAVING OUR WILDLIFE

SHADOWS OVER THE LAND

THE NATURE SLEUTHS

HOUSE SPARROWS: RAGAMUFFINS OF THE CITY

SWANS

THE HUNT FOR THE WHOOPING CRANES

THE WORLD OF THE VETERINARIAN

LORDS OF THE SKY

ANIMAL SERVANTS OF MAN

To Feed a Nation

The Story of Farming in America

by

J. J. McCoy

Thomas Nelson Inc.

NEW YORK / CAMDEN

To Claire—in fond memory of our days on the old farm near Perkiomenville

Second printing, April 1973

Published in Camden, New Jersey, by Thomas Nelson Inc. and simultaneously in Don Mills, Ontario, by Thomas Nelson & Sons (Canada) Limited.

First edition

Library of Congress Catalog Card Number: 71–145925
ISBN 0–8407–6134–1
Printed in the United States of America

Acknowledgments

The extensive research necessary to write this compact history of farming in America was made easier by the help of various persons and agencies in the agricultural field. I am grateful to the following persons and organizations for their assistance:

Zelda B. Belton, Public Services, Tuskegee Institute

E. C. Dryden, Assistant to the Director, Eastern Utilization Research and Development Division, USDA

Glenn W. Goss, Publications Director, College of Agriculture, The Pennsylvania State University

Enos B. Heisey, Field Manager, Public Relations, Agway, Inc.

Norman E. Roberts, Public Information Officer, Eastern Utilization Research and Development Division, USDA

James B. Trefethen, Director of Publications, Wildlife Management Institute

George A. Van Horn, Associate Station Editor, Agricultural Experiment Station, The Pennsylvania State University

Cecil H. Wadleigh, Director, Soil and Water Conservation Research Division, Agricultural Research Service, USDA

Agway, Inc.

Future Farmers of America

Office of Information, Photography Division, USDA

Soil Conservation Service, USDA

and the University of Oklahoma Press for permission to quote from *Plowman's Folly* by Edward Faulkner and *Deserts on the March* by Paul B. Sears.

Contents

chapter

Conservation farming in Harford County, Maryland. The Susquehanna River is seen in the background. *USDA Soil Conservation Service*

Mark, when you hear the voice of the crane
who cries year after year from the clouds
above—for she gives the signal for plowing
and shows the season of rainy winter; but
she vexes the heart of the man with no oxen.

—HESIOD, *Works and Days*

1.

The Human Food Chain

Some animals eat plants. Others eat the animals that eat plants. Man eats animals and plants, as well as their by-products. Therefore, animals and plants, obtained chiefly from livestock production and crop cultivation, are the main links in the human food chain.

Basically, a food chain is a pattern or system by which all animals (and the term *animals* includes all living creatures) are linked together by their food habits. A food chain may have a few or many links, depending upon the number of animals and plants involved in the chain.

An example of a short food chain could be a field mouse eating seeds, with the mouse, in turn, being eaten by a hawk, owl, or fox. An example of a longer food chain can be seen in the food habits of animals in or near a lake or pond. Microscopic marine animals feed on the green cells of aquatic plants in the pond or lake while they in turn are consumed by small crustaceans. Then insects eat the small crustaceans, and large fish, such as bass, trout, and pike, prey on the insects. Finally, the larger fish are taken and eaten by predatory birds, mammals, and man, thus ending this particular food chain. Man and the predatory birds and mammals are at the top of the food chain, while the small marine animals are at the bottom.

Animals are vital parts of the human food chain. But they, like man, are wholly dependent upon plants. It can be said that the basic source of all animal food is plant life. Or, to be more precise, the origin of all food is *photosynthesis* by green plants.

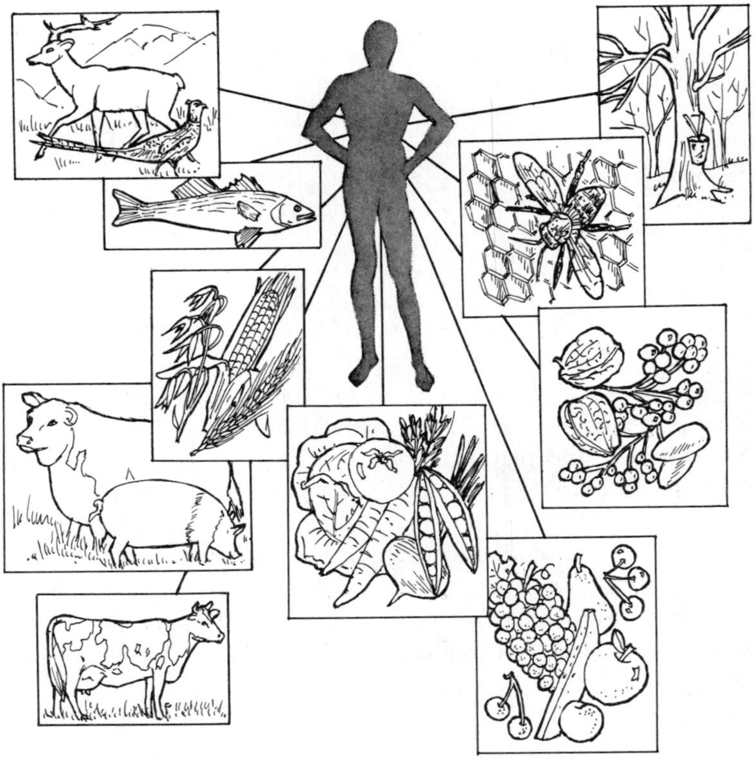

The human food chain.

Photosynthesis is the process by which carbohydrates are formed from water and carbon dioxide, through the action of sunlight on a plant's chlorophyll, or green coloring matter. Only plants have the ability to use solar energy to manufacture food. It is for this reason that the foundation of agriculture—from its beginning many thousands of years ago to the present time—is plant culture.

The human food chain is a complicated and diversified one. Man has omnivorous food habits and takes his food from various sources and in many forms. Farms, orchards, vineyards, ranches, rivers, lakes, ponds, estuaries, oceans, forests, fields, and marshes are all important sources of human food. And from these food-producing places man obtains meat, milk, fish, shellfish, fruits, berries, grains, vegetables, honey, maple sap, and ingredients for beverages.

Man is more fortunate than most animals as far as his food chain is concerned. Other animals do not exercise any control over the source of their food or the steps in its production. When a link in their food chain is broken or destroyed, the animals relying upon that link face starvation, possibly extinction as a species. Man, however, does have a greater degree of control over his food chain through crop cultivation and livestock production. By doing so, he ensures for himself a constant supply of food.

But it has taken man thousands of years to achieve this comfortable position; he has not always controlled the source and production of his food. Prehistoric man relied on the spoils of the hunt as a source of food. He not only depended upon a natural food chain like other animals, but was himself part of the food chain of large predatory mammals. The saber-toothed tiger, the cave bear, and the prehistoric wolf hunted him when other prey was scarce. Early man obtained his food when and where he could find it, mainly by hunting and fishing. Luck played an important role in this, and the caveman never knew—any more than the modern deer hunter or fisherman knows—whether he would be lucky or unlucky on a given day. Hunting skill or prowess was another important factor that determined the success or failure of a hunting expedition. Since his weapons consisted of nothing more than sharpened sticks and stones, the caveman had to be unusually clever and skillful if he wanted to get meat and hides for his family. The unskillful hunter's family went without meat and was forced to eat roots, nuts, seeds, or other plant parts that were available.

A little less than a million years ago, during what is called the Pleistocene epoch, a large portion of the earth was covered with ice. Vast regions of the Northern Hemisphere lay under huge sheets of ice; in some regions the ice was several thousand feet thick. These sheets of ice, or glaciers, advanced and receded during a period of many thousands of years. There were four major glacial periods, or movements of ice. And in between each glacial period—or roughly at intervals of a hundred thousand years—relatively warm periods occurred and melted the glaciers. The earth's landforms underwent great changes during the Ice Age. Pressed by the tremendous weight of the ice, the earth's crust subsided beneath the glaciers. Then, as the ice melted, the depressed areas rebounded. The glaciers, formed by the constant accumulation of snow, caused a lowering in the level of oceans; in some oceanic areas the level dropped as much as three hundred feet. The Ice Age was a time of turbulence and violent change for man and animals.

Prehistoric man managed to adjust to the physical changes taking place about him. Always, his main concern was to find food. He gained a stronger toehold on survival when he learned how to make better tools and weapons. When he chipped a spearhead from a stone and fastened it to the end of a stick, he found he had a weapon that was more deadly than just a sharpened stick. This new weapon gave him a better chance to bring down game, thus reducing some of the odds in the hunt. The manufacture of stone tools made it easier to cut the meat and dress skins.

Another important step in improving the food supply, or rather in making food more palatable, was the discovery of fire. It is not known just when or where early man learned to use fire. Perhaps he picked up a flaming stick ignited by an erupting volcano. Or he might have snatched up a burning branch from a tree struck by lightning. In any case, the utilization of fire was a major step in the preparation of food. Raw meat was tough and required considerable chewing to render it digestible, but when roasted over a fire, meat not only was easier to eat and digest but also tasted better.

Primitive man also needed skins and furs for clothing. At some time early in the Paleolithic, or Old Stone Age, he devised two tools that made it easier for him to process skins. One of these was the *core tool,* which was made by pounding a core or block of flint until one end had a sharp point. Hand axes were made in this way. The second tool was the *flaked tool,* made by chipping or flaking the edges of a piece of flint with another stone. Crude but effective knives and skin scrapers were made in this manner. As prehistoric man became more efficient in the art of flaking, he developed more tools. The flint blade is believed to have been developed about 35000 B.C. It was an extremely important tool, for with it early man was able to cut leather thongs from skins for use in lacing clothing.

Later, as man's culture and handicraft skills advanced, the *burin* made its appearance. The burin was a stone chisel which could cut and shape materials that were softer than stone. Deer antlers and bones were cut and carved with the burin, and so primitive man developed a whole new line of tools from antlers, horns, and bones. Eventually, he produced a sharper knife, a spearhead with barbs, and a crude bone needle.

With his improved weapons and superior intelligence, prehistoric man became the mightiest hunter on earth. His tools permitted him to cut meat and dress skins, and fire enabled him to cook his food. His weapons, his tools, and the use of fire made him much more efficient in the handling and processing of food and fiber. But he lacked control over the source of that food and fiber; he still had to depend upon the spoils of the hunt and the availability of wild plants.

The caveman's dependence upon game and wild plants as the only source of food and fiber continued for many thousands of years. But as primitive man learned to cultivate plants and tend livestock, he gained a certain amount of control over his food chain, and when man became a farmer he strengthened immeasurably his chances of survival as a species. He was no longer at the mercy of a food chain that could be snapped or cut off by the disappearance of game or wild plants. He produced his own food and fiber.

The origins of agriculture have long occupied the attention of archaeologists and anthropologists. Although prehistoric people left relics of their plant culture and evidence of livestock husbandry, scientists have not been able to point a finger at the exact area in which agriculture started thousands of years ago.

Equally interesting and speculative is the question of *how* early man became a farmer. What caused or motivated man (or rather women, for it was really the women who first cultivated plants in crude gardens) to plant seeds and tend the crops? Why did primitive man change from a hunter of animals to a keeper of animals?

There are, of course, various theories. One is that a scarcity of game forced some primitive people to eat more roots and plants, thus leading to the planting of seeds and roots. Another is that plant culture arose from an accidental planting of seeds. It may have happened that a cave woman who was hulling wheat or barley overlooked a few whole kernels, and they were tossed onto the community garbage pile. With the garbage serving both as fertilizer and soil, the seeds sprouted and grew into plants. When the cave woman realized what had happened, an idea was born: Why not plant seeds and roots near the dwellings?

The foregoing is a plausible theory. But might not the idea of simply digging a hole and planting seeds have occurred to many? Surely this is not ascribing too high a degree of intelligence to early man, who had already demonstrated in a number of ways a superior intelligence and inventiveness. And it must not be forgotten that prehistoric man had wild plants to serve as models.

Whichever way plant culture started, early man's crop patches provided him with a relatively steady source of food. As a hunter-farmer who grew part of his food supply, he now controlled a major part of his food chain, subject only to weather, disease, and insects.

Plant cultivation is believed to have originated about nine thousand years ago on the hillsides and mesas in various parts of the region known as the Near East. Roughly, this region ranged from the foothills of the Zagros Mountains on the Iran-Iraq border westward to the Jordan Valley, then north to Syria and the headwaters of the Euphrates River in Turkey, and from there into the Balkans. Two varieties of wheat, known as emmer and einkorn, were among the first plants to be cultivated in this region and formed the basis of primitive agriculture in the Near East.

As prehistoric people migrated from the Near East into India, China, Africa, and Europe, they took with them plant seeds and livestock. By planting seeds of plants grown in their original home regions as well as of the plants native to their new lands, primitive farmers expanded plant culture, which soon became the basis of ancient civilizations. Archaeological evidence shows that Old World prehistoric farmers cultivated wheat, barley, rye, flax, peas, carrots, oats, millet, soybeans, figs, dates, and citrus fruits.

Although crude when judged by modern standards, some of the prehistoric farming centers were productive. Excavations made in Iraq point to Jarmo as one of the sites where the transition from food gathering to food growing was made in prehistoric times. The carbonized remains of both wild and domesticated grains were discovered in excavations at Jarmo in 1948, along with the bones of sheep or goats, cattle, and pigs. These bones date back to 6000 B.C. (Since sheep and goats are closely related anatomically, it is difficult to tell which is which from bones, especially fossilized bones.) Thus we know that the primitive farmers living at Jarmo practiced plant cultivation and livestock production.

Early farmers at Tepe Siyalk, near Kashan on the Iranian plateau between the Caspian Sea and the Persian Gulf, developed a high level of agriculture. These farmers—actually hunter-farmers, since they still depended upon hunting and fishing as their main source of food—practiced a combination of crop and livestock farming. Their chief crops consisted of grains, such as wheat, barley, and millet. Siyalk livestock included dogs, oxen, sheep or goats, and possibly horses—although scientists are not certain as to whether the horses kept by Siyalk farmers were wild or domesticated.

Plant culture was practiced in Africa as early as 5000 B.C., and evidence of a flourishing agriculture in the region known as the Fayum has been discovered by archaeologists. The Fayumic farmers lived along desert lakes in northern Egypt. They were primarily hunters and fishermen, but they also cultivated plants and kept livestock. Excavations in the Fayum have turned up grain pits, and botanists have been able to identify some of the grains found in the pits and their approximate ages. Based on the evidence, the Fayumic farmers grew emmer wheat, two varieties of barley, and cotton. Other unidentified plants were also found. These farmers harvested their grain with flint sickles and stored it in pits lined with straw.

Excavations in the Fayum have also revealed that the primitive farmers of the region kept cattle and sheep or goats. Bones of these animals were found in several Fayumic sites.

The Indus is the longest river in south-central Asia. Early farmers cultivated plants and kept livestock in the valleys of this great river. Some of the important farming centers were in Kashmir and Pakistan, and they were equal to those in Africa and the Iraq-Iran plateaus. Farmers in the valleys of the Indus practiced irrigation on a large scale. Their main crops were wheat, barley, cotton, rice, and dates. They kept two kinds of cattle (the native zebu and a cross between zebu and cattle from the Near East), buffaloes, sheep or goats, elephants, and jungle fowl.

These are the probable origins and regions of Old World agriculture. Which region or people was the first to cultivate crops and keep livestock is really not too important, except to scholars. What is important is that prehistoric grain farmers laid the foundations for all future agricultural systems, including livestock husbandry.

And this poses another question: Did prehistoric man domesticate animals at the same time that he started to cultivate plants? Or did livestock production follow plant culture?

It is quite likely that plant culture and livestock husbandry originated at the same time, or that livestock production closely followed the cultivation of plants. One reason why this may be so is that grain production provided food not only for man but also for animals. With surplus grain and straw, prehistoric farmers could keep animals nearby. While the women cultivated plants, the men—who were hunters—would have the time to capture wild cattle, sheep, goats, horses, and pigs. And with the assurance of a food supply for the captive animals, early livestock farmers turned to livestock husbandry.

As plant culture expanded, particularly grain production,

more animals were kept in settlements, and livestock farmers became more sedentary. They built corrals in which to keep their livestock. However, some livestock farmers, notably sheep and goat herders, were nomadic; they followed their herds and flocks to grazing areas. But it can be said that prehistoric grain farming helped stabilize early livestock production.

Although livestock husbandry has always depended heavily on the harvests of crop farmers, it really is a different branch of agriculture. In fact, it is a specialized branch of agriculture, with its own particular problems. Also, crop farmers and livestock producers have different characteristics and attitudes. The livestock farmer has a greater affinity with his product, so much so, that he has often gone to war over his animals. This must have been true in prehistoric times; it is certainly true of historic times. The range wars between cattlemen and farmers in nineteenth-century America were an example of this. So were the wars between cattlemen and sheepherders. But in spite of the differences existing between livestock men and crop farmers, each depends upon the other insofar as the utilization of their products is concerned.

As more and more prehistoric people migrated from the original agricultural centers, they expanded crop culture and livestock production. Farming centers appeared in many regions of Asia and Africa, and in Europe as far north as Scandinavia. They thrived especially in regions of favorable climate.

At some time around 30000 B.C., some migrating Stone Age people ventured to leave Asia and crossed the Bering Strait by way of a land bridge to the continent of North America. Such migrations continued for about twenty thousand years. These prehistoric emigrants were hunting people who followed the wandering wild animals. It is possible that, although they did not practice farming, they were familiar with the plant culture and livestock husbandry of the Old World. In any case, their descendants in the New World were to develop a high level of agriculture far to the south of the Bering land bridge.

2.

Early American Indian Farmers

From their point of entry in Alaska, the pre-Indians spread out over North America and southward into Central and South America. They brought no livestock with them, but they may have brought some seeds. These prehistoric immigrants soon adapted themselves to a variety of climates in the Americas, from the frigid arctic tundras to the sultry tropical rain forests of South America. They settled along the coasts, on the plains, in the forests, in semiarid regions, and high in the Andes Mountains in Peru. Out of these pre-Indians evolved more than three hundred major tribal groups, with over a dozen different cultural centers of distinct patterns.

Some authorities believe that pre-Indian farming arose *in situ;* that is, it evolved in the Americas. Others think that the pre-Indians were already equipped to practice agriculture when they came to North America, ten thousand or more years ago.

Regardless of which theory is correct, the fact remains that pre-Indians developed highly successful agricultural systems. They discovered and developed crops that later became important to the existence and economy of the American Indians, the colonies, the United States, and many countries around the world. A number of plants cultivated by the pre Indians were native only to the Americas and unknown in the Old World.

Agriculture in the New World, like that of the Old World, began in several regions at the same time. It originated in regions that most favored the cultivation of plants. Since agriculture in the New World before the discovery of America reached its

American Indian farmer-hunter tribes lived in villages.

highest peak in the highlands and drier climates of Mexico, Central America, and Peru, it is reasonable to assume that plant culture had its origins in these three regions. The Aztecs in Mexico, the Mayas in Central America, and the Incas in Peru practiced crop production on an extensive scale.

It should not be supposed, however, that these prehistoric Indians started out as full-time farmers with vast fields and orchards. On the contrary, they were part-time farmers or gardeners, obtaining their food and fiber partly from hunting and fishing, and partly from their labors in the crop patches and orchards. And as in the Old World, it was the woman who cultivated the crops, while the men attended to the hunting and fishing. Agriculture among the pre-Indians (as it was with their descendants, the American Indians) was a woman's occupation and remained so for a long time. Only when crops and livestock were raised for profit did farming in America become a man's vocation.

The Old World Neolithic farmers practiced a dual type of agriculture—a combination of crops and livestock—but their New World counterparts were strictly plant cultivators. One reason for the lack of a livestock industry in the Americas before the discovery by Europeans was that there were no draft animals. Old World farmers domesticated cattle and used oxen for power. There were no cattle in the Americas, nor were there any horses. Even though the horse evolved on the North American continent, this animal had migrated to Asia long before the Old World hunters crossed the Bering Strait. Not until they were brought over by Spanish and English colonists were horses seen in the New World.

Another potential draft animal, the camel, also evolved in North America, but it, too, moved to Asia before man appeared in the New World. However, sections of the New World—mainly South America—contained animals that were related to camels. These were llamas, vicuñas, alpacas, and guanacos, which lived in the highlands of South America. Some of these camellike animals were kept by the Incas in Peru, but they were not raised on any large scale and were used as pack animals.

Yet there was a nucleus of a livestock industry in prehistoric America. In addition to the llamas, South American Indians kept guinea pigs, Muscovey ducks, and turkeys. Beekeeping was practiced by the Mayas of Yucatan, who kept honeybees in hollow logs, clay pots, and calabashes or gourds. But there were no cattle, sheep, goats, or swine.

It is believed that the absence of draft animals, such as the horse, mule, and ox, prevented early Indians from inventing and using a plow. Aztec, Inca, and Maya farmers carried on a plowless agriculture. Still, they raised crops that required heavy tillage; for example, maize, beans, potatoes, and tomatoes. Seeds and roots were planted with the aid of strong digging sticks. Ancient Peruvian farmers used a tool known as a *taclla,* a stick equipped with a foot rest or brace for pushing the digging end of the *taclla* into the soil. Other tillage tools included earth or clod crushers that resembled stone axes, and crude earth scratchers or hoes.

The list of plants cultivated by early man in America is impressive, with the plants ranging from tropical to temperate-climate species. It included maize (Indian corn), tomatoes, pumpkins, squash, gourds, melons, white potatoes, sweet potatoes, peanuts, lima beans, kidney beans, chili peppers, sunflower seeds, strawberries, avocados, Jerusalem artichokes, seedless pineapples, custard apples, cacao beans, manioc, and agave.

These plants were not only brought under cultivation by ancient Indian farmers, but were subjected to changes by selective breeding. Early Indian farmers developed special varieties of crops to adapt to specific soil and climate conditions. This plant selection must have taken place over a very long time. The early Indians also practiced a more intensified plant-selection system than any other primitive farming people. Their major crops were farther removed from the wild species than the cultivated plants in the Old World. This is another indication that the early American Indians had practiced selective plant breeding and crop cultivation for a long time.

In addition to the cultivation of manioc, a major crop, the early South American Indians devised a method for squeezing out the deadly cyanic acid from the manioc plant. It was a crude but effective press that eliminated the cyanide, but retained the valuable starch in the plant, which was used for food. (Tapioca pudding is made from the starch of the manioc or cassava plant, as it is also known.)

Prehistoric farming in the Americas evolved through four successive cultures in Mexico, Central America, and South America: the Early Farmers, the Cultists, the Experimenters, and the Master Craftsmen.

The Early Farmers were semiagriculturists, engaging in a combination of crop cultivation, hunting, and fishing as a means of obtaining food and fiber. Archaeological digs at Early Farmer sites in Peru have produced evidence that these primitive farmers cultivated achira (a plant with an edible root), chili peppers, and beans. They also raised cotton, which perhaps shows a link between Old World and New World prehistoric agriculture. Another plant grown and cultivated in both the Old and New Worlds was the gourd. The Early Farmers are believed to have domesticated llamas and guinea pigs. Anthropologists estimate that the Early Farmer culture flourished from 3000 to 1000 B.C.

After the Early Farmers came the Cultists, a slightly more advanced group of primitive farmers. They were called Cultists because their pottery decorations, usually in the forms of jaguars and ocelots, suggested a feline or cat cult. The Cultists occupied sites along the western coast of Peru and in the interior section. Their crops included maize, peanuts, manioc, and avocado and fruit trees. These early Indians are believed to have thrived from 1000 to 1 B.C.

Some advances in primitive agriculture were made by the Experimenters, a people that more or less succeeded the Early Farmers and Cultists about A.D. 1 to 600. The Experimenters followed an intricate system of farming, utilizing rivers and streams to irrigate dry lands. They cultivated new plants and kept llamas and guinea pigs. Experimenter women also preserved food, particularly meat, by drying it.

The most advanced of the prehistoric American Indian farmers were the Master Craftsmen, who followed the Experimenters. These Master Craftsmen were so called because they attained a high level of technology in comparison to previous cultures. They achieved a remarkable degree of efficiency in their farming, using irrigation canals and magnificent terraces. Master Craftsmen farmers cultivated all the plants the previous cultures had and added sweet potatoes, white potatoes, oca, seedless pineapples, an ulluco. Their crops were fertilized with fish and guano, the droppings of seabirds. Llamas, alpacas, guinea pigs, and dogs formed the basis of the Master Craftsmen livestock industry, if it can be called that.

Agriculture in the New World, like that in the Old World, began in several places at more or less the same time. The prehistoric Indian farmers passed on to their successors—the North and South American Indians—a sound system of plant cultivation. Perhaps their greatest contribution was the cultivation and development of maize, a major link in the food chain of both red and white men. Maize, or corn as it is commonly called today, became the staple food of most Indian tribes and the early colonists in North America and South America.

Corn is an important food crop throughout the world today. Some variety of corn is grown on all continents, from latitude 57 degrees north in Canada and Russia to latitude 40 degrees south in Argentina. Thousands of varieties have been developed since this unique food plant was first cultivated somewhere on the high plateaus of central or southern Mexico and the highlands of Peru. When the European explorers landed in the Western Hemisphere

North and South American Indians developed many varieties of corn. From left to right: pop corn, sweet corn, flour corn, flint corn, dent corn, and pod corn. *USDA Photo*

in the latter part of the fifteenth century, about seven hundred varieties of corn were under cultivation by Indians.

Successful corn culture requires a deep soil, ample rainfall, and constant hoeing or cultivation during its early growth. Corn is classified by six main types: pod corn (*Zea mays tunicata*), soft corn (*Zea m. amylacea*), dent corn (*Zea m. indentata*), flint corn (*Zea m. indurata*), sweet corn (*Zea m. saccharata*), and popcorn (*Zea m. everta*). This classification is based mainly on the features of the corn kernels. All these types were grown by Indians, and from them were developed varieties to suit a wide range of landforms, soils, and climates. What is most remarkable about the corn culture of the Indians is that all corn types and varieties were developed from a *single* parent species: *Zea mays*. So far, no wild species of corn has been discovered.

Although botanists have been unable to find a wild corn plant, they have found two semitropical plants that are related to corn. These are teosinte (*Euchlaena mexicana*) and a gramma grass (*Tripsacum dactyloides*), two wild plants of the southwestern United States and Mexico. Both of them have some cornlike characteristics. Gramma grass has tassels that resemble those of corn. It also bears its male flowers in the tassels, as does the corn plant. Teosinte looks more like corn than does gramma grass. It bears a branched ear, and crosses readily with corn and to a lesser degree with gramma grass. But corn and gramma grass do not crossbreed. Some botanists believe that Indian corn, gramma grass, and teosinte evolved from a common ancestor plant that grew somewhere in Mexico.

The mystery of the origin of corn persists to this day. Where did this valuable plant originate? Is it an Old World or New World plant? (Popcorn is found under cultivation by primitive people in Java, Borneo, and Burma.) And how did the early Indians come upon it and develop the numerous varieties? These questions have yet to be answered.

Regardless of the how, when, and where of ancient corn culture, the Indians must receive credit for developing this important plant. That corn was highly valued and revered by most Indians is attested to by the many legends and ceremonies that exist, from South America to the Canadian border, and from the Atlantic to the Pacific oceans. Indian corn—the *maize* of the Aztecs and the *sara* of the Incas—was grown under a system known as *milpa* farming. *Milpa* is an Aztec word for a clearing in a forest made by cutting and burning the vegetation; *milpa* farming was used chiefly for corn culture. Aztec and Inca farmers conducted their *milpa* farming with various rites, ceremonies, and prayers to the corn god, who, in the Aztec culture, was Tlaloc. Many references to Tlaloc are found in Aztec art. He is pictured on temple walls, along with corn. The Aztecs used corn tassels, ears, and pollen in their fertility ceremonies.

North American Indians extolled the virtues of the corn plant in their legends, ceremonies, and dances. Most Indian legends attribute a divine origin to corn. Iroquois myths tell about how De-hao-hwend-ji-awa-kho, the Earth Holder, came down to earth and presented the Iroquois with seeds of corn, beans, and squash. He also told them to hold a great feast and a sacred ceremony after the crops were harvested, during which they were to give thanks for the food.

Henry Wadsworth Longfellow, the nineteenth-century American poet, borrowed from Indian legends when he described how Hiawatha's people obtained corn. While Hiawatha fasted in the forest, according to Longfellow's epic poem, *The Song of Hiawatha,* he was visited by a strange youth dressed in green and yellow. Plumes of green bent down over the youth's forehead, and his hair was soft and golden. He was Mondamin, the corn youth, sent by the Earth Holder.

Mondamin challenged Hiawatha to a wrestling match. Three times they wrestled in the forest and three times the match ended in a draw. Then Mondamin told Hiawatha that he, Mondamin, would lose on the fourth trial. When this happened, Hiawatha was to strip the garments of green and yellow from Mondamin and lay him in the earth. And Hiawatha was to tend the grave, keeping it free from weeds.

Hiawatha did as he was told, and day after day he tended the grave of Mondamin—

Till at length a small green feather
From the earth shot slowly upward,
Then another and another,
And before the Summer ended
Stood the maize in all its beauty. . . .

The Indian corn stories and myths all point up the importance and impact of the corn plant on the food chain of early man in the New World. No other crop, with the exception of wheat and rice, surpasses corn in worldwide popularity, and has done more to stave off famine.

During thousands of years of cultivation, corn has evolved from a primitive plant into a highly specialized one. It grows to a height of from two to twenty feet, depending upon the type, variety, and growing conditions. The stalks from a single seed may number from one to twelve, and the leaves may range from eight to forty-eight, again depending upon the variety. The ears that bear the precious kernels vary from three to thirty-six inches in length. Corn has male and female flowers—tassels and kernels—on different parts of the same plant. This arrangement makes it relatively easy to prevent the inbreeding that usually takes place among grain plants. In crossbreeding corn, the tassels (male flowers) of one variety are removed, and a different variety is planted nearby, so that cross-pollination can take place. Ancient Indian farmers learned how to do this a long time ago and so were able to develop the different varieties of corn that the colonists found when they arrived in the New World.

With the corn plant as a key crop, agriculture spread out from its original prehistoric centers, expanding south, east, and north into the Americas. At the time of the discovery of America by Europeans, farming was an important source of food for Indians living along the eastern seaboard, although most of them still relied on hunting and fishing to supplement their food and fiber needs. The eastern tribes lived in villages, and their principal crops consisted of maize, tomatoes, tobacco, beans, melons, pumpkins, potatoes, and squash. All of the North American Indians, like those in South America and Mexico, were plowless farmers, tilling their crop patches and fields with crude diggers and hoes.

Maize, of course, was the chief crop grown by North American Indian farmers. However, some tribes living in the Great Lakes region—mainly the Sauk, Fox, Ojibwa, and Winnebago—used wild rice as their staple crop. The wild rice of the Great Lakes region is not a true rice but a plant that resembles oats. French explorers and fur traders gave it the name of *avoine folle*, or crazy oats.

A branch of northeastern Indian farming was the collection and processing of maple sap. The Indians drew the sap from sugar-maple trees, which were present in thick stands from Pennsylvania northward to Maine and westward to Ohio. Maple sugar was made in early spring, when the sap began to rise in the trees. Indian families moved into the maple woods and tapped the maples, collecting the sweet, syrupy liquid in birchbark buckets. Then the maple sap was put into wooden troughs and boiled. The heat for this came from hot stones that were dropped into the troughs. Next, the condensed sap was crystallized by being agitated with large wooden spoons. When cooled, the maple sugar was cut up into blocks and stored for use throughout the year.

To the west, the Plains Indians were divided into two distinct occupational groups: hunters and hunter-farmers. The strictly hunter tribes included the Crow, Blackfeet, Arapaho, Northern Cheyenne, Dakota, and Comanche nations. Among the hunter-farmers were the Mandan, Hidatsa, Pawnee, Omaha, and Iowa. Most of them lived in the eastern part of the Great Plains, which was less arid than the western region.

The hunter-farmer tribes were culturally related to the eastern Indians. They are usually included in the Plains Indian group because they lived as nomadic hunters of bison for part of the year. Their main crops were maize, beans, squash, and sunflower seeds, which they cultivated with hoes and other scratching tools. The Pawnee, although they still relied on hunting as a source of

The Plains Indians were mostly hunters who depended on the bison and other game for food and fiber. *USDA Photo*

food and fiber, probably were the most serious of the hunter-farmer tribes. They and other farming tribes traded their food products to the purely hunter Indians, obtaining in return buffalo meat, skins, furs, and horses.

In the Southwest, Indian farmers cultivated their crops with a high degree of efficiency in a region that had an average yearly rainfall of only thirteen inches. The Pueblos and other southwestern tribes became expert dry-land farmers. They developed a strong system of crop production by irrigating the dry lands when water was available. Since this kind of agriculture entailed heavy work—that is, digging ditches and canals—farming in the arid Southwest was a man's occupation. While the men worked in the irrigated fields, the women remained at home making pottery and other articles. Among other tribes, farming was still a woman's work.

These southwestern Indians made some important contributions to American agriculture. They developed six different colors or varieties of maize and several varieties of beans, squash, and tobacco. They were agricultural pioneers in a harsh region that later became a vast farming center of the Southwest.

Agriculture, then, had reached a high level before the coming of the white man. The early Indians had passed on to their descendants more than twenty crops and their culture. Despite the fact that American Indian civilizations and cultures were perhaps three thousand years behind those of the Old World, Indian plant cultivation provided the basis for a future giant agricultural civilization. Furthermore, the eastern woodland Indians—a Stone Age people—were largely responsible for the survival and success of the original American colonies. They rescued the starving colonists, showing them how to plant and cultivate maize, beans, potatoes, and other crops. Without these crops, it is doubtful that the colonists could have survived in the rigorous, unfamiliar environment of the New World.

3.

Farming in the Colonies

Agriculture was the way of life in the American colonies. Not all of the land in the colonies was ideal for farming. New England soils were rocky and difficult to till; therefore many of the people in this region turned to the sea for a living and became fishermen. But there were fertile, workable soils and more favorable climates in the middle and southern colonies. Large and small farms appeared in these regions, and the farmers tilled the soil, cultivated crops, tended livestock, and themselves built their houses, barns, and other buildings. In time, farmers in the middle and southern colonies became nearly self-sustaining, raising almost all of their food and fiber.

In New York, Pennsylvania, and Maryland, where the soil was deep and fertile, large crops of wheat were grown for home use as well as for shipment to England. Farther south, in the Virginia Tidewater area, in the Carolinas and Georgia, colonial farmers raised indigo, tobacco, cotton, and rice. All of the colonies, including the rugged, bleak New England settlements, practiced some kind of agriculture, even if the produce was intended only for home or local use.

Once the colonists had ensured their permanence in the New World by producing enough food and fiber for their own needs, they set about expanding their agriculture. This meant placing more land under cultivation. Land was plentiful; the wilderness seemed endless, and all a farmer had to do was clear away the trees for more cropland. And this the colonists proceeded to do with great energy and determination. Forests toppled

Early American colonists owed their survival to the Indians, who showed them how to plant corn and other crops.

to the bite of countless axes, swamps were drained or filled in, barns were erected in an astonishingly short time by neighbors joining in a barn-raising party, new homesteads were built with the help of neighbors, and livestock corrals, sheds, and pens were fashioned from split logs as more and more farms appeared in the colonies.

Colonial agriculture flourished, and soon the farmers found themselves with a surplus of food and fiber. They began exporting the surplus to England, along with raw materials such as lumber and naval stores. The exportation of farm products and raw materials to England fitted into the mercantile theory of trade that was the doctrine of England and other major European powers. It was their belief that wealth was measured by the amount of gold and silver in the treasuries. Therefore, the powerful nations developed a system of trade that was designed to bring the precious metals into their coffers.

The American colonies played an important role in England's balance-of-trade system. It was vital that England obtain food, fiber, and raw materials from her colonies. If she could have prosperous, self-sustaining colonies to buy her manufactured goods, money would flow into and remain in her treasury. Equally important, this exchange of goods and money would have the added effect of making the mother country less dependent on foreign nations, which could never be entirely trusted, and might turn into enemies at any time. The colonies would be a secure source of food and raw materials in times of war as well as peace.

With this idea in mind, each of the great maritime nations—Spain, England, France, Portugal, and Holland—established and maintained colonies in the New World. The colonies could trade only with the mother country; no other country could trade with them. Thus, only the parent country would benefit from the exchange of food, fiber, raw materials, and manufactured goods.

England forbade her colonies from developing any kind of industry that would compete with British manufacturers. This restriction caused the colonies to concentrate on the development and expansion of agriculture and the production of raw materials, but it could not deter some American colonists from engaging in a manufacturing business. New England colonists did start to produce goods, but a strong manufacturing system was not developed for a long time.

There were other good reasons why manufacturing was slow to develop in the colonies—aside from the ban by England. Land was cheap and plentiful, and this fact made farming more attractive and easier to develop. Many of the colonists came from farmer stock; few had any knowledge or skill in manufacturing. There was a scarcity of labor for any large-scale manufacturing industry. Also, machinery for mills and factories was very expensive and had to be transported from abroad. All of these deterrents argued for agriculture as the basic industry of the fast-growing colonies.

England's insistence on the mercantile system eventually became one of the chief causes of the American Revolution. A one-sided trade that was restricted to England severely limited both the agricultural and industrial development of the colonies. However, a kind of trade truce existed between England and the colonies for the first sixty years of the eighteenth century. During this period, the colonies expanded and improved their agriculture and small industry. They became prosperous and more self-reliant. But there were no large cities to compare with those in England and Europe. The American colonies were populated by farmers, shopkeepers, laborers, and some wealthy merchants and planters.

Colonial America was a rural land, with the majority of the colonists occupied in farming. There was a lot of work on the farms. The women fed and clothed the family, and the men and boys cleared land, planted crops, and tended livestock. When the harvest of a large crop of wheat, corn, or oats needed extra labor, a farmer's neighbors turned out to help him bring in the crop. Although horses and oxen were used as draft animals, much of the work on a colonial farm was done by hand and with crude implements.

Eventually the balance of trade went against the colonies. They soon owed more money to England than their agricultural and industrial products were worth. By the middle of the eighteenth century, the colonies were indebted to England at the rate of £200,000 a year. Many colonial leaders believed that this indebtedness could be wiped out if the colonies were permitted to trade with all of Europe. But the English Parliament refused to remove the restrictions on colonial trade.

Money became scarcer and scarcer in the colonies. This situation forced the colonists to make local payments with various commodities, or what the colonists called country pay. In the southern colonies, for example, teachers and preachers often received their wages in tobacco. Pennsylvania and New York farmers paid for services with wheat certificates. At Harvard College, students who came from farms often paid their tuition with farm products. As a result of the money shortage, some New England assemblies listed a number of farm products and other commodities that could be used by colonists to pay their taxes.

These and other problems piled up on the colonists. Out of the differences with England, plus a desire to be free, came the American Revolution. The Revolutionary War and its military actions did not cause any large-scale expansion in crop or livestock production. Most farms continued to produce their regular harvests. The American army could not boast of being the best fed and best clothed in the world, despite a flourishing agriculture. On the contrary, Washington's army was poorly clothed and often went without food, especially when the British or their Indian allies, the Iroquois, captured farms, burned crops, and slaughtered livestock.

Agricultural expansion took place after the Revolutionary War. It was obvious to the leaders that if the fledgling United States was to take a place among the major nations of the world, it would be necessary to expand agriculture and industry. The new nation was still a country composed of small cities and large farm communities. Its agriculture was on a firm footing, but some of the older croplands were beginning to show signs of unproductivity. Farmers had followed a crude and exhaustive kind of crop culture, with few or no attempts made to restore soil fertility. Northern farmers now sought new means of increasing the productivity of their farmlands. In the South, the planters looked around for new lands to bring under cultivation.

Agricultural and political leaders recognized the need to expand their new country's crop and livestock production. The best way to do so was to put more land under cultivation. But the colonies, now states, were already bursting at the seams with people, towns, farms, mills, and shops. Beyond the Allegheny Mountains, however, lay a vast region that could provide the necessary lands for agricultural expansion. But the lands between the Alleghenies and the Mississippi River were claimed by eight states: Connecticut, Georgia, Massachusetts, New Hampshire, New York, North Carolina, South Carolina, and Virginia. The claims on lands west of the Alleghenies, which permitted western expansion, were based on the original charters of these states.

Five states—Delaware, Maryland, New Jersey, Pennsylvania, and Rhode Island—had no claims on the western lands. Their boundaries had been fixed by their original charters. This situation put them at a disadvantage; they had no land with which to reward their Revolutionary War veterans and thus encourage expansion.

Maryland, one of the states with a fixed boundary, insisted that those states with western holdings should relinquish their claims to the United States. And to lend force to her argument, Maryland held back ratification of the Articles of Confederation for three years, until this matter was settled. Finally, the eight states ceded their western holdings to the United States. With the acquisition of the western lands, the new nation could now expand physically, agriculturally, and industrially. New states could be created out of the western lands. It was the start of the phenomenal growth of the United States.

Into the lands across the Alleghenies streamed war veterans, farmers, and other Americans who wanted cheap or free lands. They were a rugged breed of men, half farmer and half hunter. They traveled over the mountains on foot and horseback, fighting Indians and the wilderness. Their hunting prowess came in handy and kept them alive until their farms were established—until the first crops had been harvested and livestock were producing meat, milk, eggs, and skins. Ten years after the end of the Revolutionary War, more than a hundred thousand of these hardy settlers were tilling the rich soil in the valleys of the Ohio and Cumberland rivers after conquering the raw wilderness.

After the Revolutionary War, agriculture expanded in the number of farms, the amount of acreage under cultivation, and the variety of crops produced. But it still lagged far behind European agriculture in techniques and practices. American farmers, industrious though they were, still cultivated their crops and raised livestock with crude implements and methods. Little was known about soils, fertilizers, plant and animal breeding, insect control, plant and livestock diseases, and other important phases of farming. Very little agricultural information was available to the farmers, and what there was generally turned out to be extremely limited in content and distribution. Europe, however, enjoyed large resources of agricultural information and had several agricultural and veterinary schools. America had none of these advantages.

Despite the lack of agricultural information and farm and veterinary schools, America had a number of men who were learned in the natural sciences. Many of these men—botanists, zoologists, and geologists among them—encouraged agricultural improvement by their works and writings in eighteenth-century America.

Benjamin Smith Barton, a professor of natural history and botany at Philadelphia College, wrote a book called *Elements of Botany*, which covered the cultivation and growth of agricultural plants. Another scientist, Gerard Troost—a Dutch physician, chemist, and geologist—was the founder and first president of the Philadelphia Academy of Natural Sciences. Troost contributed papers on agriculture to the newly formed Philadelphia Society for Promoting Agriculture. He was especially interested in the relationship of chemistry and geology to soils.

John Bartram, the celebrated Quaker botanist who established America's first botanical garden in Philadelphia, had a wide interest in agriculture. He not only discovered new plants but conducted experiments on plant culture and farming techniques. Bartram and his neighbors cleared land along the west bank of the Schuylkill River and by a series of embankments protected the land from being flooded. He also farmed part of his lands, raising various crops for home and local use. A special feature of Bartram's garden was the orchard, which contained native and imported fruit trees. Farmers and scientists from other states came to Bartram's garden to examine his horticultural exhibits and observe his farming practices. Many of them returned home with valuable information gained from the gentle Quaker scientist.

Owners of large plantations, such as George Washington and Thomas Jefferson, were vitally interested in improving American agriculture. Both men owned vast plantations on which they conducted experiments in crop cultivation, fruit culture, and livestock production. Each of these nationally prominent men shared his farming knowledge with other farmers and planters.

Washington did not follow the usual farming practices—which were geared toward simply producing a crop with no attempt at maintaining soil fertility—but farmed with a view toward upgrading farming conditions for himself and other farmers. He began farming with tobacco as a cash crop, then switched to the production of wheat, forage crops, and livestock. Neighbors often observed Washington's agricultural experiments at Mount Vernon, which included the cultivation of wheat, clover, timothy, and vegetables. Although information on soil fertility was scarce, Washington tried to improve the productivity of his soils by applying gypsum, marl, and salt to his fields. He also improved his farm implements and invented a seed drill.

A progressive farmer, Washington read whatever agricultural books he could get from abroad, and made detailed notes on what he read. Among the farm books owned by Washington were Tull's *Horse Hoeing Husbandry* and Home's *The Gentleman Farmer.* He carried on an extensive correspondence with Arthur Young, England's leading agricultural authority and editor of *The Annals of Agriculture.*

The other great Virginia statesman and planter, Thomas Jefferson, had a more scientific and inventive mind than George Washington. The master of Monticello owned more than ten thousand acres of land, which were planted in wheat, corn, clover, field peas, and potatoes. He had some very ambitious plans to restore and revitalize his fields, as well as definite ideas about the improvement of agriculture in America. He strongly disapproved of absentee farm ownership (a number of plantation owners did not live on their lands), believing that a farmer or planter should live on his lands and be in close touch with his crops, livestock, and field hands. Jefferson also warned other farmers about planting crops that impoverished the land, and he advocated crop rotation and stressed the use of the best farm implements.

"The most important of all questions a farmer has to decide," wrote Jefferson, "is crop rotation." In keeping with this dictum, he set up a seven-year crop-rotation system on his plantations, using wheat, corn, legumes, and potatoes as the rotation crops. Jefferson had a strong aversion to growing tobacco, not only because of the moral question (Jefferson considered the use of tobacco as sinful), but because tobacco rapidly impoverished the soil. He favored the cultivation of wheat, but did not care for corn as a major crop. However, since corn was a basic staple food in the ration of his slaves, he had to include it in his crop rotations.

Jefferson, like Washington, was an avid reader of agricultural books. He also corresponded with Arthur Young and wrote long passages about his agricultural experiments. A *Garden Book* and *Farm Book*—two journals which he faithfully kept and in which he jotted down notes and observations on farming—were Jefferson's written contributions to agriculture and were the most complete records of any American of the period.

As a man interested in improving agriculture, Jefferson became a leader in the movement against the staple or single-crop agriculture practiced in the South. He believed in a diversified, self-sufficient kind of farming, such as that practiced by northern farmers, and he constantly urged his neighbors to take care of the soil and improve their farming techniques. One of Jefferson's major contributions to agriculture was his improvement of the moldboard plow. (The moldboard is that part of the plow just behind the share or cutting point. It receives the sliced furrow and turns it.) Jefferson's plow was designed according to mathematical principles; it was constructed to offer the least resistance in turning the sod, and so did the job better than the ordinary moldboard of the time. But Jefferson's greatest contribution to

American agriculture was in stimulating farmers and planters to take pride in their profession and to improve their farming methods.

Another gentleman farmer who strongly believed in the future of America and her agriculture was Michel Guillaume Jean de Crèvecoeur, a Frenchman who lived on a farm in New York. De Crèvecoeur, who wrote under the pen name of J. Hector St. John, traveled through northern and southern states, visiting farmers and planters. He spent some time with John Bartram in Philadelphia, examining the famous botanical garden and farm. De Crèvecoeur also visited Dr. George Logan of Stenton, just north of Philadelphia, a man Jefferson considered to be the best farmer in Pennsylvania.

De Crèvecoeur recorded his travels and observations in a book, *Letters from an American Farmer,* which was published in 1782. The letters were written to friends in Europe and told about the people, towns, and farms of young America, portraying life on the farms and plantations in great detail. Although primarily a literary work, the book was an important addition to America's slowly growing source of agricultural information. De Crèvecoeur also introduced some European crops, such as alfalfa, into America.

Although men like Washington, Jefferson, Bartram, Troost, and de Crèvecoeur were eager to learn about, and pass on, agricultural information, the majority of American farmers possessed little technical knowledge. They farmed by methods handed down to them, methods that were often inefficient and sometimes injurious to the land. Many of them could not read, and even if a mass of agricultural information had been available they could not have made use of it. Not until a public school system was established did farmers and other rural people receive training in reading, writing, and arithmetic.

In the latter part of the eighteenth century, the farmer's knowledge of agriculture was greatly increased by the formation of agricultural societies. These societies were, for many years, the chief source of farming information. The Philadelphia Society for

Promoting Agriculture, an outgrowth of Ben Franklin's American Philosophical Society, was the first one to be founded. Its object was the promotion of "a greater increase of the products of the land within the United States." This society published papers dealing with farming, offered prizes for agricultural essays, and made awards for farm experiments and improvements. It did not limit membership to farmers and agriculturists in Philadelphia or Pennsylvania but welcomed members from all thirteen states. Among its more illustrious members were George Washington, Thomas Jefferson, Robert Livingston, and Noah Webster.

The society urged the Pennsylvania legislature to incorporate a state society for the promotion of agriculture. It also proposed that American boys should be taught agriculture as a part of their regular schooling. Along with this suggestion, the society recommended that county agricultural societies be formed, with schoolmasters as secretaries. With the aid of agricultural textbooks, the schoolmasters could teach agriculture along with other subjects. But the society's proposals went unheeded by the Pennsylvania legislature.

More agricultural societies soon appeared. South Carolina established a Society for Promoting Agriculture and Improving Other Rural Concerns. Another society was founded in Hallowell, Massachusetts, which at that time was part of Maine. This society held meetings for the reading of papers on farming practices. It also served as a social club for farmers and their families.

In New Jersey, the Morris County Society for Promoting Agriculture and Domestic Manufacture was established, along with a library. The New York legislature granted funds to Columbia University for a professorship that would include agricultural instruction. Samuel Latham Mitchell, a distinguished physician, was appointed Professor of Natural History, Chemistry, and Agriculture. His courses contained lectures on soils, manure, plant diseases, and foods. Dr. Mitchell was active in promoting agriculture throughout New York and wrote papers on grasses, cankerworms, and other topics of interest to farmers.

Near the end of the eighteenth century, the Massachusetts

Society for Promotion of Agriculture was established with a state charter. This society became extremely active in amassing and distributing agricultural information to local farmers. It published farming articles in newspapers and in pamphlet form, created a chair of natural history at Harvard University, and founded a botanical garden. The Massachusetts society also introduced new seeds and plants, promoted the development of improved livestock, held cattle shows, and in general brought new ideas to the attention of farmers of Massachusetts.

An agricultural society formed at Wallingford, Connecticut, was to provide the basis for later agricultural experiment stations. An announcement of the Connecticut society stated: "An object of investigation for the Society shall be agriculture, with such subjects of inquiry as may tend to explain its principles." The Connecticut society conducted agricultural experiments and field observations and distributed its findings to the farmers of Connecticut.

A few agricultural books appeared toward the end of the eighteenth century. The first distinctive American farm book to be published was *Essays Upon Field Husbandry in New England,* by Jared Eliot. Samuel Deare of Massachusetts produced *The New England Farmer,* an encyclopedia of agriculture, and J. B. Bardley of Maryland, a judge, wrote *Essays and Notes on Husbandry and Rural Affairs.*

Perhaps the most significant, practical contribution to American agriculture at this time—one that revolutionized southern farming, at least—was Eli Whitney's invention of the cotton gin. Previously, cotton lint had to be separated from the large cotton seed by hand, a time-consuming and laborious process. Whitney's cotton gin separated the lint a hundred times faster than hand labor could. It removed the fibers from the cotton seed by means of spikes projecting through slots into a chamber that held a mass of bulk cotton. The spikes, working like mechanical fingers, pulled the cotton lint through the slots, which were too narrow to admit the cotton seeds. A revolving brush cleaned the cotton lint from the spikes.

Whitney, a Yale graduate, later made some improvements that increased the cotton gin's efficiency. This mechanical cotton separator enabled planters to harvest cotton in good weather and, if necessary, gin or clean it during bad. And it set the stage for the development of the fantastic cotton empire that was soon to be created in the South.

As the eighteenth century drew to a close, American agriculture was productive and in the process of expansion. Equally important, it was on its way toward becoming a science and the lifeline of the young nation.

4.

The Pioneer Farmers

When Thomas Jefferson instructed Robert Livingston and James Monroe to negotiate the purchase of the Louisiana Territory, he had two objectives in mind: (1) to prevent the control of the west bank of the Mississippi River from being transferred from Spain (which had just ceded the Louisiana Territory to France) to the hands of the Napoleonic government, and (2) to acquire more land for the United States.

The Louisiana Territory was one of the most valuable tracts of real estate in the world. It embraced a vast region of North America that extended from the Gulf of Mexico to the Canadian border, and from the Mississippi River westward to the towering Rocky Mountains. It was a wilderness of mixed landforms, climates, waterways, and untold natural resources. The acquisition of this immense region more than doubled the area of the United States upon the mere stroke of a pen and the expenditure of $14.5 million.

Jefferson's giant real estate venture was to provide lands for raising cattle, sheep, horses, swine, and mules—thousands of acres for rippling wheat, vast stretches of waving corn, and sugar and cotton plantations in the southern part. The Louisiana Territory was to help make the United States the agricultural giant of the world. A little more than a hundred years after the purchase of the territory, the value of its farm properties was nearly $34 billion—more than two thousand times the original purchase price.

President Jefferson sent Meriwether Lewis and William Clark

Wagon trains of emigrant farmers rolled over rough roads and wilderness trails to the rich farmlands to the west.

to explore a great part of the new territory. They set out in May of 1804, and their famous expedition was in the field for more than two years. Lewis and Clark traveled through some of the most scenic and fertile country in North America. Their wilderness trek started at Wood River just above St. Louis, and took the explorers through Illinois, Missouri, Kansas, South Dakota, North Dakota, Montana, Idaho, Washington, and Oregon.

Jefferson issued detailed instructions to Lewis, the senior captain of the expedition, as to the object of their explorations and some of the results he expected. "The object of your mission," wrote Jefferson, "is to explore the Missouri River, and such principal streams of it, as—by its course and communication with the waters of the Pacific Ocean, whether the Columbia, Oregon, Colorado, or any other river—may offer the most direct and practicable water-communication across the continent, for the purpose of commerce."

Lewis and Clark were also instructed to familiarize themselves with the various Indian tribes and to learn as much about them as possible, particularly about their agriculture and their hunting and fishing, and their culture.

Lewis and Clark opened the door to the West and through that door went a steady flow of adventurers, hunters, trappers, and mineral seekers. They were rough, brawling, and restless, and more intent on adventure, excitement, and quick wealth than on settling and developing the new addition to the United States. After them came farmers, town-builders, industrialists, and merchants—the more stable people who were prepared to spend their lives in developing the fantastic resources of the Louisiana Territory. They were ready and willing to challenge the wilderness, risking raids by hostile Indians, and battling the harsh weather of the northern regions of the territory.

The pioneer farmers, the graybacks and sodbusters, swarmed into the territory. Lands immediately across the Mississippi River were the first to feel the bite of the ax and the gouging of the plow. Trees were chopped down, lands cleared, and virgin prairie soil was churned up to make seedbeds for corn, wheat, oats, barley, and other crops. Teams of oxen strained to yank out stumps and labored in the hot sun as they pulled clumsy wood-and-iron plows. Men and boys sweated as they pried up stumps, cut and notched logs, or guided bouncing plows through the deep prairie sod. Roughhewn cabins, sod huts, split-log barns and sheds soon appeared on what once had been open prairie.

They worked hard and long, these hardy pioneer farmers. Their cleared lands were plowed from sunup to sundown. Slow-moving oxen or long-eared mules provided most of the brute force, but at the end of the day the pioneers often fell into bed with sore muscles and aching backs from the hard work. The work of the pioneer women was never done; they labored constantly to feed and clothe their men and children.

Life was rough on the frontier farms. Farmers had to provide food and shelter for their families and livestock, trusting in luck and praying to God that the land would be kind to them. But the land was not always kind; there were hot summers that dried the soil and blistered the skin, and icy winters that froze the body and numbed the mind. In his struggles to conquer the wilderness, subdue the hostile Indians, and just to stay alive, the pioneer farmer became self-reliant, resourceful, and courageous. He often lived at great distances from towns and even neighbors. He had to do for himself and family, leaning heavily on his own ingenuity and skills. In the process, the western farmer became part of a new society or group that had its own special needs.

Early pioneer farms were not much to look at. They were crude layouts that provided just the bare essentials for man and beast. The fields were laid out in grids or squares. Boys walked along and dropped five or six kernels of corn at the spot where the grid lines intersected, usually about two or three feet in either direction. A man with a hoe followed behind the boys and covered the seed with soil. At harvest time, the corn was cut with a sickle and stacked in shocks. In cold weather, or at special husking bees, the husks were stripped from the ears of corn. A husking bee was a merry event, a time of fun and relaxation. Boys and girls participated, and whoever found a red ear was rewarded with a kiss.

Wheat and other grains were harvested with cradles; these were scythes equipped with a wooden framework to catch the falling stalks of wheat without shattering the grain. Men moved in a line over a wheat field, rhythmically swinging their cradles as they cut down the golden wheat stalks. Women followed the men, bending down to pick up bundles of wheat and tie them into sheaves. After threshing, the grain was stored in granaries near the barn.

Hay was important as forage for livestock. Timothy and other grasses were cut with scythes, then scattered for drying with wooden forks. When the hay was cured, farmers raked it into windrows with wooden hand rakes that were six feet wide, or with horse-drawn flop rakes. A farmer walked behind the flop rake and triggered a release that dumped hay accumulated by the tines of the rake. Then the piles of hay were heaped into haycocks with forks or brought directly into the barns.

Ten years after Lewis and Clark pointed out the path to the great West, the Louisiana Territory was well on its way to becoming a rich farming region. Agricultural expansion in the territory was furthered by three events. The first was General William Henry Harrison's victory over the Indian tribes under the leadership of the Shawnee chief Tecumseh at the Battle of Tippecanoe. This victory was important to western farmers because it reduced the threat of Indian raids on farms. The second factor was the completion of the Cumberland Road—a stone road extending from Cumberland, Maryland, to Wheeling, West Virginia, and later continued through Indiana and Illinois. It provided an easier way to the West by allowing emigrating farmers to make connections with the Ohio and Mississippi rivers. The third major development that helped western agricultural expansion was the putting into operation of the first steamboat on the Mississippi River. This steamboat replaced the old scows and flatboats, which had to be poled against the current of the mighty Mississippi, and floated with the current. Now transportation to the West was much quicker and more comfortable on the Mississippi steamboat.

With the establishment of farms and the increase in population, the western territories developed quickly. In the period between 1810 and 1821, four new states were added to the Union: Indiana, Illinois, Missouri, and Mississippi, all of them agricultural states. Now, more than a quarter of the population of the United States lived beyond the Allegheny Mountains and across the Mississippi River.

Western farmers began to broaden their agricultural markets to include the East and South. Corn, oats, wheat, bacon, and beef were their specialties, and these were eagerly purchased by people east of the Alleghenies. Although the Cumberland Road was a major artery to and from the West, many farmers found it difficult to transport their products to distant markets. Produce had to be hauled in heavy wagons pulled by horses or mules. Tributary roads were rough, often pitted with deep holes or lined with ruts and furrows. Heavy spring rains or thaws eroded the roads or completely washed them out. Speed was impossible on the wilderness roads; some of them were so bad that it did not pay a farmer to risk taking his products to eastern markets. Road improvements were not easy and they were costly. Most western farmers had little cash to pay for road repairs or new roads. They just managed to meet the payments on their lands and farming implements.

Nevertheless, the western farmers were now making a significant contribution to the food and fiber needs of America. They could no longer be ignored, or regarded as people who had received cheap lands and therefore should make out as best they could. The pioneer farmers were a growing and integral part of America's agricultural and industrial economy and were therefore in a position to make demands on the government.

Nationalists such as Henry Clay and John C. Calhoun believed that the United States should consolidate the various sections of the fast-growing country. They argued that America should also formulate its own national policy, a strictly American policy that would not lean on Europe and in which the American farmers and industrial workers would be dependent on each other and no one else. Under this policy the factories would create a market of industrial consumers who would purchase farm products and in turn supply the farmers with the manufactured goods they needed.

This seemed like a practical idea, but putting it into action was something else again. Sectionalism had already set in. The interests of the three major sections of the United States—North, South, and West—produced rivalry instead of cooperation, discord instead of harmony. Industrialists in the North opposed cheap land policies that attracted men to the West, thus draining the eastern labor supply and raising wage levels. Other northerners were unhappy with the increasing power of the western states and territories in Congress, a power that threatened the supremacy of the East.

The South did little to include the western states and territories as part of the agricultural-industrial complex of the United States. True, the South purchased a large portion of its food from western farmers, but it offered little in return. Southern cotton went to northern mills and to England. The South also had no interest in seeing the West connected with the North by a system of roads and canals, especially when it would have to help pay the costs of such facilities.

As a result of the increasing sectionalism, western farmers found themselves in a predicament. They had performed a remarkable feat in increasing their agricultural output, and under extremely difficult circumstances. But their extraordinary productivity had backfired on them: they were unable to consume all that they produced. And although they had this surplus of farm products, western farm regions lacked the money to attract and develop an industrial population that could have consumed the surplus food.

Faced with this kind of situation—a gradually hardening sectionalism—the nationalism advocated by Clay and Calhoun was doomed to failure. Sectionalism simply turned it aside. Each year, sectionalism increased, sharply dividing the country and bringing it closer to civil war.

But although sectionalism created problems for western farmers, American agriculture as a whole was steadily advancing as a profession and science. Crop and livestock production moved upward. Southern planters experienced new peaks in cotton production. Eli Whitney's cotton gin made the production of cotton highly profitable. Its use had sent cotton export figures higher and higher every year. Cotton yields rose from 2,000,000 pounds in 1791 to nearly 200,000,000 pounds in 1821.

Another factor that made cotton a profitable crop was the use of slave labor. With Whitney's cotton gin and slaves, the South developed the plantation system of agriculture, a system which continued until the Civil War.

But the emphasis on cotton and tobacco—the single-crop farming of the South—soon brought difficulties to southern planters. Cotton and tobacco quickly exhausted the soil. Few planters bothered to revitalize their soils; in fact, little was known about soil and its requirements. Southern cotton and tobacco fields were not retired from production to rebuild their structure and fertility. They were simply abandoned. The demand for cotton and tobacco was so great and the profits so high that southern planters just used their fields until they were worn out. Then the planters obtained new fields.

This kind of farming was a ruinous one. It was also a wanton waste of a vital natural resource: the soil. And it caused more land speculation. After Andrew Jackson had cleared the Mississippi Territory of Indians, southern planters occupied this fertile region. This meant that more than a third of the nation's cotton crop was raised in lands west of the Alleghenies.

In 1833, Edmund Ruffin of Virginia started a monthly publication called the *Farmer's Register,* a 64-page magazine about farming. Ruffin, a self-educated book farmer, was interested in soil fertility. He had already received some attention for his theories and experiments on the causes of soil sterility. He believed that acid wastes from vegetables and other plants were the primary cause of soil sterility, and that the application of marl or lime to soils would neutralize the acids and convert them into manure that would help restore soil fertility.

Ruffin conducted various experiments on his farm and kept extensive records on soil analysis, the use of marl, vegetable manuring (the plowing under of vegetable matter, today known as green manuring), crop culture, and farm production costs. He talked and wrote about soil bacteria, something that few people understood or accepted. Unfortunately, Ruffin was looked upon as a crackpot by most farmers of the day, and his ideas about the use of marl or lime were ridiculed.

But Ruffin's farm magazine became an instant success with southern farmers who were experiencing difficulties with their worn-out soils. This pioneer soil chemist preached about crop rotation, proper plowing technique, use of animal and vegetable manure, farm drainage, and the economic use of slave labor. Southern farmers soon realized that Ruffin wrote about his own personal experience on his farm, that he was not just a theorist.

Unhappily for Ruffin and the farmers, he branched out into politics and campaigned against the bankers, whom he labeled as enemies of agriculture. His attacks on bankers soon led to pressures from Virginia politicians, who considered his views as dangerous. Subscriptions to his farm magazine were canceled in large numbers, and the loss of revenue forced him to abandon publication. The death of the *Farmer's Register* was not only a blow to Ruffin but to many southern farmers who had profited from the agricultural information in the magazine.

Since the South concentrated on single-crop production, it did little in the way of manufacturing. Actually, the plantation system excluded the development of any manufacturing industry in the South. There was no labor force to work in the factories; most of the slaves were used in the cotton and tobacco fields. Since there was no industry in the South, the planters were almost wholly dependent on the North and West for the supplies which they could not produce themselves. This condition was to cause the South serious problems in the Civil War.

Under the stern guidance of Brigham Young, who succeeded to the leadership of the Mormon Church after the murder of the founder, Joseph Smith, wagon trains carrying more than fifteen hundred Mormon farmers and craftsmen headed toward the Far West. Driven from their city, Nauvoo, and surrounding farms in Illinois, the Mormons traveled across the prairies and deserts. They reached Utah, which they called Deseret, and settled near the Great Salt Lake.

The Mormons were industrious and skilled farmers. Even though most of the land in which they settled was not as fertile as the prairies to the east, Mormon farmers, by hard work and perseverance, developed a great agricultural center in Utah. Inter-

Mormon farmers and craftsmen, led by their leader, Brigham Young, headed west to Utah, the land they called Deseret.

spersed with barren, arid lands were areas of relatively high fertility, and these the Mormons proceeded to turn into productive croplands through irrigation. Their main crops were wheat, oats, corn, barley, buckwheat, flax, hemp, and fruits. They were excellent livestock husbandmen, raising quality cattle, horses, mules, sheep, and swine.

America had never seen quicker, more active, industrious, and contented settlers and farmers than the Mormons, or Latter-Day Saints. Their skill and zeal were unsurpassed in the history of western settlement and the development of American agriculture. Mormon contributions to agriculture were considerable, and their farming added to the growing productivity of the western territories.

A mechanical revolution took place in agriculture during the first half of the nineteenth century. All kinds of labor-saving farm machinery made an appearance. John Lane started to manufacture all-steel plows, which were an improvement over the old wood-and-iron plows. Cyrus McCormick invented and manufactured the first practical reaper, which cut six acres of oats in a day. McCormick was born on a farm near Lexington, Virginia, and became interested in farm implements at an early age. He applied himself to the improvement of farm machines and tools. McCormick foresaw the great grain-producing potential of the Midwest and established a farm machinery factory in Chicago.

More farm machines were developed and put into use. Patents were issued for a two-row corn planter, a mowing machine, a disk plow, a grain combine, a threshing machine, a wooden hand pump (which used suction to lift water on the farms), a portable steam engine, and the first hay baler. McCormick's reaper and other new farm machines were cumbersome and often cranky in operation. Nevertheless, they were the first products of an American technological and mechanical skill that would ultimately produce a fantastic multitude of farm machines and equipment.

The great increase in farmlands and therefore in crop production, and the development of labor-saving machines, called for a more readily available supply of agricultural information, and soon newspapers, magazines, journals, and textbooks were being published all over the country. Government sponsorship and distribution of farm information, however, were still nonexistent, at least on an official basis.

One government agency, the United States Patent Office, did compile and distribute farm information. The Patent Office, without any authorization, collected and published agricultural statistics. Henry Leavitt Ellsworth, a lawyer, farmer, and businessman from Connecticut, was the director of the Patent Office. Under his leadership, an agricultural division of the Patent Office was created for the purpose of collecting, collating, compiling, and distributing agricultural statistics. Ellsworth also obtained seeds and plants from consular services and naval personnel and distributed these seeds and plants to interested farmers. Two years after Ellsworth had been appointed director of the Patent Office, he convinced the government to appropriate funds for agricultural information. And although the initial appropriation was very modest, it enabled the Patent Office to increase its efforts in collecting and publishing agricultural statistics, conducting agricultural investigations, and distributing seeds and plants to farmers.

Despite the fact that the Patent Office made some important contributions and represented the only government agency providing a service to farmers, it became clear to many agriculturists that it was not the ideal source for agricultural information. But since it was the only federal farm-information center, it had to suffice. Nothing was done to remedy this situation until the second half of the nineteenth century.

Another major source of agricultural information for farmers were the lyceums, or public lectures. These lectures or meetings broadened the field of agricultural education and also laid the foundation for a national educational system. The lyceum movement in America was organized by Josiah Holbrook, who came from a farm near Derby, Connecticut. Holbrook first established an industrial school in 1819; five years later, he founded an agricultural school. Unfortunately, his agricultural school was unsuccessful, largely because of a lack of funds.

After the failure of his farm school, Holbrook turned his attention to the lyceum or lecture idea. The first lyceum was held in Millbury, Massachusetts, and instantly became a success. A large number of people turned out to hear the lectures. Basically, the lyceum movement provided literary and scientific knowledge to adults who were not enrolled in formal schools or colleges.

By 1831, there were lyceums in nine hundred towns, at which farmers and other interested persons could hear lectures on farming and other topics. A few years later, more than three thousand communities had active lyceums. One lyceum, the American Lyceum—which was not a national organization—was instrumental in improving public schools and founding libraries and museums. Holbrook dreamed of a national lyceum, but this dream never materialized. Lyceums remained community programs.

As American agriculture expanded and developed, the need for more technical information and education increased. It was obvious to educational leaders and agriculturists that if American agriculture was to be improved, it would need a continuous supply of knowledgeable farmers, and that it was necessary, therefore, to provide a technical farming education to the sons of farmers. This meant that agriculture would have to be taught in rural schools, and that more agricultural colleges should be established. It also meant that America's flagging educational system had to be revived and strengthened. In the latter part of the 1830's, general education was weak and inefficient, mainly because of ignorance and prejudice on the part of local school boards.

To correct this situation, Horace Mann, a Massachusetts educator, launched a great education revival program in 1837. Mann was the first secretary of the Massachusetts Board of Education, which he had helped to found. He played a key role in setting up a tax-supported educational system in Massachusetts and worked to improve education all over the United States.

Mann visited schools in Europe, particularly in Prussia, where he was impressed with teacher training and general classroom procedures. Upon his return to the United States, he campaigned for educational reforms. He organized training institutes for teachers and brought about the establishment of three normal schools, forerunners of teachers' colleges.

American agricultural and vocational schools grew out of the work of Philipp Emanuel von Fellenberg, a Swiss educator, agriculturist, and philanthropist. Fellenberg founded a vocational school at Hofwyl, Switzerland. At this school, students worked three hours a day at agriculture, horticulture, and mechanical arts. The rest of the day was devoted to the study of academic subjects. Fellenberg demonstrated that the labor of teachers and students on the school farm and in the workshops could make the school and students partly or wholly self-supporting.

A few manual-labor schools appeared in the United States, mostly in the Northeast. Some of these vocational schools taught only agriculture, others combined agriculture with mechanical arts. They provided students with practical work in the barns and fields, along with the regular academic subjects. The quantity and quality of the agricultural courses taught in these schools varied considerably. Most of the courses were superficial, only skimming the surface of agricultural technology. In a sense, these early farm schools were premature, for a sufficient store of agricultural knowledge suitable for use in elementary and high schools had not yet been developed. Still, these schools were important; they stimulated a movement that was to culminate in the establishment of agricultural colleges and experiment stations.

One of the early farm schools was the Boston Asylum and Farm School located on Thompson's Island in Boston Harbor. The founders of this school stated the purpose as follows: "The establishment of a farm school in the country—where idle and morally exposed children of the city may be rescued from vice and danger, and may enjoy the advantage of a good physical and moral education—would not only be a great benefit to such children, but would greatly conduce the peace and good order of this community."

Such an avowed purpose for a school would hardly attract youngsters of today. But then, the students of the Boston Asylum and Farm School had little choice in the matter. They were not volunteers but were sent to the school right from the streets of Boston.

Children attended school for six hours. They were taught the elements of reading, writing, grammar, arithmetic, and geography. After the academic sessions, they received instruction in agriculture, horticulture, and botany. They were also given practical training in soils, crop cultivation, and care of fruit trees. In addition to the formal instruction, the students worked on the school farm.

Most of the manual-labor and farm schools were owned and operated by private organizations, such as churches and missions.

Although these schools were helping to expand and extend agricultural education and information, they were inadequate for the task. Agriculturists urged state aid for agricultural education. More than that, they began an active campaign to establish state agricultural schools and colleges.

Edward Hitchcock of Deerfield, Massachusetts, president of Amherst College, was interested in agricultural education. He went to Europe to survey the farm schools in Germany, France, and Switzerland. Hitchcock conducted extensive research on these schools, concentrating on how they were operated, what they taught, and how they were financed. He concluded that American agricultural education—as well as farming practices—was retarded when compared to European methods. He also stated, on his return to the United States, that agricultural schools would be the best way to introduce agricultural improvements to farmers.

By the end of the first fifty years of the nineteenth century, American agriculture was experiencing growing pains. There were more farms, increased crop cultivation, more livestock, new machines, a few agricultural schools, and a slowly accumulating store of information. All of these advances had been made by the industry, dedication, and determination of farmers, rather than by the application of science and technology. American agriculture still had some distance to go before it would catch up with that of the more progressive European countries.

But American farmers were attracting the attention of Europe. Their grain yields, livestock production, and flow of new farm machines forecast an agricultural revolution. With continued industry and the acquisition of technical knowledge, the future of American agriculture in the second half of the nineteenth century appeared bright. And even though American farming techniques and practices lacked the science and technology of those of Europe, America's potential as a giant agricultural country was visible among the flaws.

5.

Agriculture on the Eve of the Civil War

The United States census for 1860 showed that American agriculture was flourishing in almost every area. Sixty percent of the working force, out of a population of more than thirty million people, was engaged in farming. Food and fiber production climbed steadily; one farm worker produced enough food to feed himself and five other persons. Improved farming practices, aided by the advent of a variety of machines and equipment, increased production on many farms.

Europeans were becoming more aware of the agricultural revolution that was taking place in America, or at least of its mechanical phases. The London Exhibition of 1851 featured American farm machinery. Reapers, seed drills, threshers, and other time- and labor-saving machines made in America proved that the United States had not only the newest machines, but was in a position to supply other countries. In fact, some American machines were already in use in England and the European continent.

It was the steadily increasing annual production of crops, particularly corn, wheat, hay, and cotton, which served to indicate that American agriculture was now moving into the world scene. The progress of American agriculture amazed many observers at home and abroad. They stated that the remarkable advance of American agriculture—made through applied science and use of machinery—was more rapid than that of any other agricultural people. In practically every department of farming, mechanical power had wrought a revolution.

The greatest contributions of mechanical skill and power to improved farming techniques came from tillage, harvesting, and processing machines. Grain farmers now relied heavily on the reaper, thresher, winnowing machine, and other harvesters. The commercial importance of the wheat crop alone made it necessary for farmers to get their crops into markets at the earliest time and in the best possible condition.

In just about twenty years, the increase in wheat grown in all the states and territories amounted to 70 percent. This was double the increase rate of the population in the same period. The greatest gain in wheat production was, of course, in the midwestern states and territories. Older wheat-producing states, such as Pennsylvania, New York, Ohio, and Virginia, did not show a proportionate increase in production.

Indian corn continued to be a major crop. The total corn crop on the eve of the Civil War amounted to more than 830 million bushels, an increase of over 40 percent in a twenty-year period. Unquestionably the most popular crop, corn was grown in every state and territory. Farmers found corn to be a crop that was less liable to failure than any other. Also, corn had a variety of uses as human and animal food. There were no important changes made in either the varieties of corn or cultivation methods, except the substitution of machines for human and animal labor.

Thirteen states were producing the nation's cotton crop as the clouds of civil war grew heavier on the horizon. More than five million bales of cotton were being exported abroad, mostly to English mills. Cotton planters found themselves in an enviable position: The demand for their cotton now exceeded the supply, and they could look forward to continuing growth in the production of their specialized crop.

The extraordinary success of the cotton crop and demand for it led southern planters to believe that cotton was king—the number-one crop of the United States. To support their claim, they pointed to the fact that, in 1850, nearly 50 percent of the nation's exports was in cotton, as opposed to 9 percent in manu-

Southern planters depended on slave labor to harvest cotton and other crops.

factures and 7 percent in flour, beef, and pork. Since cotton was produced with slave labor, the South maintained that slavery was the basis of the country's prosperity.

Dairy farming was concentrated in the northeastern and middle states and some of the territories. Raw milk was in great demand. The manufacture of cheese and butter increased over a twenty-year period, but cheese did not reflect as high a rate of increase as did some other dairy products. Although cheese was a nourishing food and well suited for inclusion in the workingman's lunchbox, it was not a popular daily food of the average American. In contrast, European working people consumed large quantities of cheese as part of their daily food ration.

There were two main reasons why cheese did not figure prominently in the diet of the average American. First, cheese manufacture was not yet a fully developed art in the United States. Very little of the cheese made could be called first-rate. Second, cheese was expensive and not many workingmen could afford to include a slice or slab of cheese in their lunchboxes. To stimulate the consumption of cheese, dairy farmers and manufacturers sought ways to improve the cheese-making industry.

Sugarcane was grown in the extreme southern states, but the crop was far short of the country's needs. Western states and territories had difficulty in obtaining enough sugar. This situation was partially solved by the introduction into some western states of a sugar-producing plant known as shaloo or sugargrass (*Sorghum saccaratum*), a plant native to China and Africa. The success of western farmers in growing shaloo indicated that this plant might become an important addition to American crop production. At least, the shaloo grown by some western states was making them less dependent on the scarcer and more expensive cane sugar of the South.

Tobacco had a more fantastic rate of increase than any other crop; the total amount produced in 1859 was 430 million pounds, an increase of about 106 percent in a ten-year period. This was an amazing growth rate for a crop that was known to impoverish the soil and was suspected of harming its user. Yet the demand for tobacco created ready markets and good profits for the southern planters producing this crop.

Livestock production also showed an increase, particularly in the western states and territories. Hogs, beef cattle, sheep, and poultry were bringing profits to livestock farmers in all regions, except the South, where crop cultivation was the chief agricultural occupation. In the West and Southwest, hardy longhorn cattle were displacing their cousins, the bison, on the rangelands. Mormons were now firmly entrenched in the territory they called Deseret and were raising superior beef cattle, horses, and sheep. Flocks of sheep were now becoming more common in the West. More than sixty million pounds of wool reached the market in 1860 from both eastern and western wool growers.

America's rangelands were ideal for sheep production. Only

two other regions in the world had lands equally well suited for this purpose: Australia and Argentina. But despite the excellence of the western rangelands, sheep farming was developing more slowly than other branches of the livestock industry. The main reason for this was that cattlemen objected to the presence of sheep on the rangelands.

Sheep were competitive with cattle for grasses and other grazing plants. The sheep, or woollies, as they were called, were also destructive to the grazing lands, for they cropped the grasses too close, injuring the plants and leaving little or no grazing for cattle. Also, cattlemen claimed that they had priority on the rangelands, since they had pioneered the region and fought off the hostile Indians. These and other arguments provoked bitter feuds between cattlemen and sheepherders. Violence flared up, and there was open war between the two opposing groups of livestockmen. Men and animals on both sides were killed as the feuds became more widespread.

Horses and mules supplied power for farm and city. Draft-animal production was a major part of the livestock industry and was centered in the Midwest. Work animals had to be fed forage, and hay production was an important industry in the northern and middle states. The annual hay crop now amounted to nearly twenty million tons, which was not enough for the nation's needs. There was room for improvement in hay production, and agriculturists believed that the hay crop could be increased by the introduction of better varieties of grasses, rather than by increasing the acreage.

One area in which there was little progress was that of treating animal diseases. There were no veterinary schools in the United States. Most of the animal doctors, regardless of whether they called themselves veterinarians, farriers, or horse doctors, were men with little or no scientific training or education. A few legitimate veterinarians were practicing in the country, but these were mostly European immigrants who had received a veterinary education abroad.

The scarcity of qualified veterinarians resulted in high rates

A sheepherder's wagon is surrounded by his flock on the range. Sheepherding in the West was a lonely job. *USDA Photo*

for livestock insurance. Livestock diseases went unchecked and took a high toll of animal life. These two factors—high insurance rates and high mortality from diseases—prevented the growth of the livestock industry in the more isolated parts of the country.

Hog cholera, pleuropneumonia, and cattle fever continued to cause a high mortality. More than $25 million worth of hogs died every year from cholera. Pleuropneumonia, a disease introduced into America by a cow purchased from a British sea captain by Peter Gunn, a New York milkman, claimed the lives of thousands of cattle every year. Beef cattle ranchers fought a losing battle against cattle or Texas fever. Sheep raising was difficult in some regions because of a disease known as scabies and because of serious infestations of intestinal parasites.

Livestock farmers called on the federal government for help

in reducing animal deaths from these diseases. But there was not as yet any government agency for this kind of assistance; furthermore, not much was known about the causes, treatment, and control of animal diseases in America.

Another problem that plagued farmers in all sections of the country except the arid regions was the drainage of wet fields. The problem was more acute in those regions where flooding occurred each year. Tile drainage systems were being installed in wet fields and these promised to correct the condition. Tile factories began to produce special tiles for farm drainage systems, and formerly neglected farmlands were being put into crop production. Unfortunately, tile drainage systems were expensive, usually beyond the means of the average dirt farmer.

Irrigation of western lands, mainly in Arizona, California, New Mexico, and Utah, was also bringing more lands under cultivation. Irrigation farming was not a new idea of course; the Incas and some other ancient Indian civilizations had constructed some very effective irrigation systems. Now, in the years just preceding the Civil War, the reclamation of lands by irrigation was receiving more attention. It was a farming system that showed great promise for regions with little rainfall; a system that could turn thousands of acres of barren ground into fields of luxuriant vegetation.

The United States Agricultural Society urged Congress to establish a federal agriculture department. Although there was some interest on the part of some senators and representatives, no action was taken on this recommendation. Farmers still had to rely on other sources for information and advice on crop cultivation and livestock production.

Only five states had agricultural schools: New York, Pennsylvania, Michigan, Maryland, and Iowa. Instruction at these institutions was mostly on the high school level. In fact, the agricultural school in Pennsylvania was known as the Pennsylvania Farmers' High School. These farm schools were struggling for survival and their total effect on the nation's agriculture was negligible.

Justin Smith Morrill, a congressional representative from Vermont, introduced a resolution that called for a look into "the expediency of establishing one or more national agricultural schools on the basis of the naval and military schools" (Annapolis and West Point). Morrill's resolution stipulated that one scholar from each congressional district and two from each state at large would receive a scientific and practical agricultural education at public expense. Congress did not accept this resolution.

Undaunted by the rejection of his resolution, Morrill persisted in his efforts to establish more agricultural schools. He, like others interested in the welfare and advancement of agricultural education, believed that the country needed more farm-oriented schools. The son of a Vermont blacksmith, Morrill had been a farmer and a merchant before he became a politician. In the House of Representatives, he served on the Committee for Territories, then became a member of the Agriculture Committee.

Morrill sponsored another agricultural education bill. Briefly, this second bill authorized the transfer of public lands to the states and territories. These lands were to be sold and the proceeds used to establish and maintain colleges for agriculture and mechanical arts. Approximately 6,340,000 acres of public land—held in trust for the people by the United States Government—were to be apportioned to each state and territory. Each state would receive twenty thousand acres for each senator and congressman representing that state. Delaware, for example, received 90,000 acres; New York was given 900,000,000 acres to sell.

An important feature of the Morrill College Land Grant Bill, as the proposal became known, was that the federal government would, at least indirectly, give support to the nation's agriculture. Except for the limited work of the Patent Office, federal support of agriculture did not exist. Many agriculturists believed that the lack of federal support was actually holding back the improvement of important branches of agriculture, such as livestock husbandry. Agricultural research was scant and disorganized. The new role of American farmers—that of world food provider—was not an easy one. If they were to meet their obligations efficiently, then it was necessary to supply them with the latest scientific information and techniques in crop and livestock production.

The Morrill College Land Grant Bill was a step in the right direction to provide the basis of scientific agriculture. These colleges could conduct research as well as train future farmers. But Morrill's plan had many opponents who claimed that it was unconstitutional. Firmly established liberal arts colleges saw the proposed agricultural and mechanical arts colleges as threats. However, supporters of the bill pointed out that the liberal arts colleges need not fear the agricultural and mechanical schools as potential competitors. Farmers and mechanics required special training that was not available in the regular colleges. Agricultural and mechanical arts students should, of course, receive an education in the liberal arts, but such courses could be designed to meet the needs of the students. Therefore, argued the friends of the Morrill plan, agricultural and mechanical colleges would not be in conflict with liberal arts colleges.

As for the charge that Morrill's bill was unconstitutional, he and other supporters of the measure pointed out that Congress had found a way to promote commerce, educate officers of the Army and Navy, and grant lands and other favors to the railroads. Surely, Congress could do the same for agriculture, the nation's main industry.

The Morrill bill was subjected to changes, additions, and deletions in an attempt to make it acceptable to everyone. Among the opponents of the bill were Senators Clay of Alabama, Green of Missouri, and Jefferson Davis of Mississippi. These men argued that the bill violated the right of the states to manage their own domestic affairs. Others questioned the authority and right of the federal government to dispose of public lands in such a fashion. However, the right of the government to do so was plain, absolute, and unlimited. After all, the government had given lands to soldiers, settlers, railroads, new states and territories, and schools and colleges. Certainly it had the right to do the same for agriculture.

Finally, the Morrill College Land Grant bill was sent to President James Buchanan for his signature. Buchanan, a man with roots in the soil, was a former member of the House Agriculture Committee. Under pressure from various groups and individuals who opposed the bill, however, Buchanan returned the measure to the House of Representatives with a veto message. Morrill and his supporters were stunned by the President's action; they had thought the revised bill would be accepted by all.

Buchanan summed up his objections to the bill in his veto message. He stated that the bill would deprive the United States Treasury of almost five million dollars, which the sale of the public lands would bring in during the next fiscal year. The bill encouraged the states to rely on the federal government for aid to which they were not entitled. Buchanan, obviously influenced by colleges and universities, stated that the bill was unjust in that it would interfere with, and probably injure, colleges already established with private funds. He claimed that the bill would harm new states, because it would force down the value of land scrip (a paper issued as evidence that the holder was entitled to receive an allotment of land) and make it possible for land speculators to acquire large tracts of land for resale. Finally, the President stated that the bill was unconstitutional because there was no grant of power to the federal government to use public money for the benefit of people in the states and territories.

Friends of agriculture refused to accept defeat on the Morrill College Land Grant Bill. It was too important a measure, one that affected the advancement of agriculture, to let die. The United States Agricultural Society, along with Morrill and other important leaders, kept the issue alive and reaffirmed their belief in the measure whenever they had the opportunity. A presidential election was near, and the two major candidates, Abraham Lincoln and Stephen Douglas, were approached by agriculturists as to their intentions about the Morrill College Land Grant Bill. Both Lincoln and Douglas indicated that they would sign the bill if elected.

President Lincoln did sign the bill into law in 1862, thus establishing the land-grant college system that was to become a keystone of agriculture in the twentieth century. In the same year, Lincoln approved the formation of the United States Department of Agriculture, appointing Isaac Newton, a prominent Pennsylvania dairy farmer, as the first commissioner of agriculture. Newton was a staunch friend of agriculture and had served as superintendent of the agricultural division of the Patent Office. He was also a member of the influential United States Agricultural Society.

But now the Civil War engulfed the nation in flames and hatred. The Morrill Act and other needs of agriculture were set aside for the more urgent task of saving the Union. Sectionalism had triumphed. The nation was torn in half, South against North. And agriculture, especially in the South, did not escape the devastation of civil war.

6.

Reconstruction Agriculture

The Civil War brought prosperity to northern and western farmers; disaster and impoverishment to many southern planters. In the North, dairy farming, poultry production, fruit culture, and market vegetable gardening reached a high peak of production. Western grain and livestock farmers greatly increased their yields with the aid of animal power and improved machinery. There was an all-out effort to produce food and fiber for civilian and military needs.

In the war-torn South, the situation was quite different. Food and fiber production was at a bare maintenance level. Most able-bodied white men were in the army fighting for the Confederacy. Although the slaves performed most of the field work, there was a scarcity of competent overseers or farm managers to supervise the work of the slaves. An effective northern blockade prevented the exportation of cotton and tobacco, the South's two great cash crops, to European markets. Horses and mules were commandeered for use in the Confederate Army. Much farm machinery was ruined. Many large plantations were destroyed when northern officers, acting under orders from General William Tecumseh Sherman, put the torch to crops and buildings in the campaign to break southern resistance.

Various issues and problems arose at the end of the Civil War. Two major questions were: How should the United States use its victory over the rebellious South? And what was the new status of the southern states—were they still part of the Union, or had they forfeited their rights as states and should therefore be treated as territories?

President Lincoln believed that the southern states were still part of the Union. He had declared in a proclamation of December, 1863, that when 10 percent of the voters in any of the seceded states would form a loyal government and accept the laws of Congress—and his proclamation as to slavery (the Emancipation Proclamation)—he would recognize any such government as legal. Unfortunately, Lincoln's offer met with considerable opposition in Congress. And unhappily for the South, he did not live to push his program for restoring the solidarity of the almost-shattered Union.

After the war, thousands of blacks, mostly field hands, suddenly found themselves free men. Many of them thought the

Tobacco, one of the South's main cash crops, helped to restore agriculture almost destroyed by the Civil War. This photo shows tobacco cutting in West Virginia. *USDA Photo*

Day of Jubilee had arrived. But the facts proved otherwise. Few of the blacks had any training for work other than farming; none had any money with which to start a business or buy a farm. A large number of them had been led to believe that the plantations of their former owners were to be carved up and divided among them; thus, every former slave would receive forty acres and a mule.

To complicate matters, unscrupulous politicians from the North, the carpetbaggers, further confused the blacks with all kinds of wild promises. The carpetbaggers posed as the friends and benefactors of the freed blacks. In reality, they were simply working for their own profit and political advancement.

Everything was at a standstill at the end of the war. Southern agriculture had been dealt a mortal blow. Farmers and planters—those who were lucky enough not to have lost their farms and plantations—were disorganized and dispirited. The future seemed hopeless. This seemingly hopeless plight of the southern planters during the humiliating Reconstruction period has formed the background of a number of novels about the South. Stark Young's *So Red the Roses* and Margaret Mitchell's *Gone With the Wind* are two famous books that depict the deplorable conditions in the South during Reconstruction. Neither of these books exaggerates the extent of the ruined plantations or the despair felt by the defeated South.

It was clear that if the South was ever to regain its economic feet, something had to be done about restoring its agriculture. At first, an attempt to continue the old plantation system was made, since the system had proved so effective and profitable before the war. Cotton and tobacco were still in great demand, and prices were high. But the revival of the plantation system failed. Now the South's labor pattern was different; gone was the abundant and cheap slave labor of antebellum days. Emancipated blacks were not accustomed to a wage system, and they often would not work continuously or efficiently.

With difficulty, a large cotton crop was produced during Reconstruction, so much that a surplus occurred, which in turn caused a drop in prices. Debts incurred by planters in reestablishing their plantations mounted, and many southerners were in desperate financial straits. Their situation became more aggravated by the excessively high taxes levied by the carpetbaggers. Some planters were forced out of business, and their lands were either sold or left idle. The cash value of southern plantations and farms declined 48 percent between 1860 and 1870.

As a result of the financial squeeze on the large planters, their holdings were cut up, divided, and parceled out in small tracts to whites and some blacks. Most of the blacks became tenant farmers, working for a share of the crops; some whites also accepted this kind of life.

Gradually and painfully, southern farms and plantations returned to the nation's agricultural production line. More fertilizers and better machinery helped to speed up production. The cotton yield increased from 172 pounds per acre in 1860 to more than 200 pounds ten years later. There was some diversified farming in the South—vegetable gardening, fruit culture, dairying, and livestock production—in an attempt to produce food. But the bulk of southern food and other supplies was purchased from northern and western farms. Machinery and other farm equipment also came from the North and West. A credit system was set up, but this only added to the troubles of farmers and planters, for it kept hundreds of them, especially blacks, in financial bondage.

Elsewhere in the nation, agriculture was undergoing some changes. Most of the industrious and financially stable northern farmers lived in orderly communities. They were well adjusted to the financial, industrial, and social complexities of the times. But the poorer northern farmers were not adjusted or satisfied with their conditions. Their poverty and discontent caused them to look westward, where they hoped to acquire cheap or free lands and start new farms and rural communities. An exodus from the North to the West took many farmers away from their native farms. Impoverished southern farmers and war veterans also left their home states for what they hoped would be a better life in the West.

The Midwest had its restless farmers, too. Some pioneer farmers looked for greener pastures in the Far West. A number of them sold their farms to the newcomers from the North and South and moved on to California, Oregon, and Washington. This constant population agitation and movement to the West characterized the post-Civil War period. Men and women were determined to find a new life, new outlets for their skills. Some of them were willing to settle down on new farms and work hard. Others hoped to amass a fortune by luck or get-rich-quick schemes. The West was considered a land of great fortunes, the El Dorado of America, with a promise of better things for everyone. Unfortunately, for many the West turned out to be as unkind as the East or South had been.

Still, the West did offer opportunities for those who wanted to farm. Even if a farmer was just a field hand or laborer, he could do better in the West. He could take advantage of the Homestead Act of 1862 and settle on government land. After a few years, he would become the owner of 160 acres of reasonably good farmland. Many immigrants did take advantage of this particular opportunity.

Railroad construction and expansion also promised better opportunities in the West. The railroads were the lifelines of the western farmers, providing them with a means of shipping their farm products to eastern markets and ports. In the post-Civil War period, the railroads were the friends and allies of farmers who supplied the nation's food and fiber.

The rapid growth of the railroads after the Civil War was a bench mark of the country's prosperity. But more railroads and tracks were needed to link the various regions of the United States. The Civil War had shown the need for a better and faster connection between the states and farm belts. Therefore, efforts were turned toward speeding up and facilitating the expansion of the railroads.

In 1862, President Lincoln had signed a bill granting permission to the Union Pacific Railroad to build a line from the western edge of Iowa to a point where it would connect with a road to be constructed by the Central Pacific Railroad. The CPR was to lay tracks eastward from Sacramento, California. Large parcels of land were given to both railroads, lands that would take the tracks through Indian territory and wildlife habitats.

Private investors were wary about this transcontinental railroad system, despite the backing and generosity of the federal government. They were not convinced that the UPR and CPR tracks could be linked up. There were about eighteen hundred miles of wilderness to be spanned, from the Missouri River to the Pacific Ocean. The tracks had to be laid through hostile Indian country. They had to travel through oven-hot deserts and over freezing mountains. Many persons were skeptical as to the practicality of the project; others said the plan was doomed to failure.

But the railroads did the job. Ties and rails were laid by gangs of rough, quarrelsome, hard-drinking Irishmen and enigmatic, slow-moving Chinese laborers who sweated and cursed in the deserts and froze in the mountains. Shiny steel tracks were spiked down on heavy ties to the chant of "Working on the Railroad," roared out by the labor gangs. On May 10, 1869, the Union Pacific rails met those of the Central Pacific at Ogden, Utah. Railroad officials and other dignitaries pounded spikes of gold and silver into the last ties. Then the locomotives of both railroad lines approached each other and touched noses, as some of the spectators said afterwards.

Two months after the last spike had been driven, a train sped eastward from California. Six and a half days later, it arrived in the Jersey City station in New Jersey. The link between East and West was completed. It meant the end of the overland stages and slow freight wagons. From this day onward, it would be the Iron Horse that raced across America, hauling passengers, manufactured goods, and farm products.

The great westward migrations, begun when Lewis and Clark

blazed the trail, and stimulated more than sixty years later by the extension of the railroads, changed the character of the lands beyond the Mississippi River. As more lands were opened for settlement, farming, mining, and timbering, the ecology of vast regions was drastically altered or destroyed. Natural plant and animal communitities were knocked out of balance by the encroaching civilization with its farms, mines, mills, lumber operations, railroads, and people.

Western wildlife, particularly the large grazing mammals, was endangered by agricultural expansion. More than sixty million bison once roamed the short and tall grass prairies, from Mexico to the Canadian border and from the Alleghenies to the Rocky Mountains. Millions of nimble pronghorn antelope had also bounded over the prairies. After the extension of the railroads and the arrival of farmers, ranchers, and town builders, the herds of bison and antelope started down the trail toward extinction or greatly reduced populations.

Thousands of bison and antelope were killed to feed railroad gangs and soldiers. Buffalo tongues were shipped east to fancy restaurants, and the hides were sent to the furriers. Bison and antelope were competitive with cattle and sheep for the rangelands; they were enemies of agriculture. Cattlemen and sheepherders encouraged the extermination of the bison and antelope by contract hunters, such as Buffalo Bill Cody. In an astonishingly short time, the bison population was reduced to an alarming low point, one that pushed this mammal close to extinction. By 1880, less than a thousand bison remained out of the great herds that once had thundered over the prairies. The antelope were also decimated to token populations in the western states and territories. Other wildlife faced the threat of the advancing civilization: whooping cranes, trumpeter swans, grizzly bears, mountain lions, wolves, and coyotes were shot, trapped, or poisoned to the point of extinction.

But more than wildlife was disturbed or destroyed by the postwar development of the West. The region was the home of a number of Indian tribes that depended on the bison, antelope,

Nebraska settlers pose in front of their sod-roofed home in 1887.
USDA Photo

and various wildlife for food, clothing, and other commodities. Some of these tribes were hunting people, others were farmer-hunters, but all of them relied on the availability of game. The hunter tribes were the most seriously affected by the destruction of game and loss of lands. The Northern Cheyennes, Dakotas, Comanches, Blackfeet, Apache, Utes, and Nez Perces were gradually pushed out of their ancestral lands, but not without considerable bloodshed. The Indians bitterly and actively resisted the eviction. Many times they went on the warpath. They raided farms and ranches, killing settlers, slaughtering livestock, and burning crops and buildings.

It took twenty years of fighting—at a cost of over twenty million dollars—before the Indians were subdued and the West made safe for farmers and ranchers. When the bison and antelope were nearly exterminated, and the Indians reduced to poverty and inactivity, western agriculture, especially the livestock industry, started another great advance.

Immigration from Europe helped both the expansion of agriculture and the growth of America. Germans, Swedes, Norwegians, Danes, and Poles came to the United States after the Civil War. Many of them were farmers who went west, attracted by the promise of cheap or free land. Those states and territories which encouraged immigration and accepted the new people made notable advances in their agriculture, industry, and arts. The infusion of immigrants into the hardy pioneer farmer stock of the West produced an industrious agricultural work force. It was a combination that was to help make America the agricultural giant of the world.

The first ten years after the Civil War were good years for farmers in most regions except the South. They enjoyed an agricultural boom, with high yields and high prices. But the boom came to an end. During the war, demands for America's grains and livestock were high and farmers prospered. Ten years later, domestic and foreign markets began to slacken off. Foreign demands for America's farm products diminished because European agriculture was improving also. Western grain farmers received a major setback when bumper crops of wheat produced in Russia and Argentina dropped the price of wheat by nearly a third of the prevailing price.

Trouble was brewing for the rest of the country, too. The nation's capital had been invested in various ventures which brought in little or no returns. To remedy this situation, the federal government tried to put money on a sounder basis by reducing the paper currency. This action, coupled with restrictions on the coinage of silver, brought about a deflation of the economy. Two large and highly destructive fires in Boston and Chicago caused extensive property damage. These two tragedies added to the country's growing financial difficulties. Farm prices and wages were lowered. Everyone was affected: farmers, ranchers, industrialists, government, and the ordinary workingman. The overall result was the Panic of 1873.

As farm prices kept falling, farmers began to voice their grievances. They protested against lowered prices, high cost of freight on railroads, and the reduction in their buying power. They blamed all of their troubles on two factors: (1) excessive rates charged by the railroads and grain elevators or storage units, and (2) the substitution of gold for cheap greenback currency. They pointed to the fact that they were getting ninety cents for a bushel of wheat which sold in eastern markets for one dollar and fifty cents. More and more farmers found themselves in financial difficulty as the cost-price noose tightened around them. In the Midwest, some farmers were burning corn on the cob because at fifteen cents a bushel it was cheaper fuel than coal or wood.

Even though farmers formed a distinct group or segment of American society and industry, they had no national guilds or organizations to represent them. There were agricultural societies, but these were organized on a state or local basis and were concerned more with accumulating and distributing farm information than with obtaining better prices for farmers.

Factory workers and other laborers were moving toward unification as a working group. But not the farmers. They remained a paradox: they formed the largest group of workers and producers in America, yet had little or no voice in regulating the economy of agricultural production. Farmers were strong individualists, and fiercely independent. Individualism and independence had been keystones in the remarkable success of pioneer farmers, but they were now millstones around the necks of the post-Civil War farmers.

America's farmers needed some kind of organization that would bring them together and increase their influence and bargaining power. One man did dream up a farmers' organization. He was Oliver H. Kelley, a clerk in the Department of Agriculture. Kelley came up with the idea of a secret farm organization, with a ritual and different orders somewhat like those of the Masons, of which he was a member.

Kelley and three other agriculturists had founded the National Grange of the Patrons of Husbandry in 1867. William Saunders, a well-known wheat specialist, was elected Master of the Grange, with Kelley installed as secretary. Kelley stated that the main object of the Grange was "to encourage and advance education in all branches of agriculture." He later broadened the objectives of the Grange to include an increased farm production by the application of science.

These were high-sounding words. But Kelley and the other founders of the Grange, who certainly must have been aware of the practical needs of farmers, omitted the most important ingredient in a farmers' organization: protection against high costs of production and fair prices for farm products. They soon found out, when they tried to recruit members for the Grange, that the average dirt farmer wanted actions or measures to protect him from the injustices of the railroads, granary operators, and commission merchants.

Grange leaders realized that their organization had some serious shortcomings and they took steps to correct them. The objectives of the Grange were revised so as "to advance education, to elevate and dignify the occupation of the farmer, and *to protect its members against the numerous combinations by which their interests are injuriously affected.*"

With protection written into its constitution, the Grange quickly caught on with farmers in most sections of the country. Soon there were more than a thousand local granges, of which more than six hundred were located in Iowa, the heart of the grain-producing region. Thousands of farmers joined the Grange, and all sections of the country eventually had local granges; there were also ten state granges.

The conditions brought about by the Panic of 1873 intensified and widened the discontent of the farmers. Protests and calls for action were common in rural areas. Since the Grange seemed to be the only organization interested in the farmer's plight, its membership multiplied enormously in a relatively short time, reaching a peak of one million in 1875.

Grange leaders set about to find ways to help distressed farmers. They organized farm cooperatives through which farm products were sold and supplies purchased, eliminating the middleman. Farm machinery and equipment was manufactured by some cooperatives, but this ran into legal troubles when private manufacturers claimed infringement of patents. The Grange also established banks and life insurance companies. Many farmers received business training and learned about the powers of organization through the local granges.

Since the railroads were part of the farmers' troubles, the Granger Movement, as it was called, campaigned for laws regulating the railroads. As a result of Granger pressure, several states did enact laws that regulated railroads and grain-elevator rates. But the railroads retaliated by rendering poor service and, in some cases, by actually mishandling shipments of grain and other farm products. Most of the courts were unsympathetic toward the so-called Granger Laws and declared them to be unconstitutional.

The Grange was an important adjunct to American agriculture, a movement which increased the influence and bargaining power of the farmers. Although it encountered considerable opposition and resistance, the Grange accomplished some major reforms. For one thing, it awakened farmers and the general public to some of the inequities of farm production and marketing. It secured the passage of laws for the establishment of state railroad commissions, taxation reforms, advancement in public education, creation of state boards of agriculture, and prevention of pooling by railroads. Pooling was a device used by railroads to keep freight rates up. In pooling, shipping was divided among the various railroads that would ordinarily compete for it. The total profits were then placed into a common treasury or pool and divided according to the business assigned to each railroad. It was a device used to eliminate competition between the railroads and, of course, to prevent the lowering of shipping rates.

Between 1870 and 1880, a million acres of semidesert lands were made tillable through irrigation. The Desert Land Act of 1877 stimulated the reclamation of arid lands. Under this act, settlers were offered 640 acres at $1.25 per acre, provided they would irrigate the lands. Many farmers took advantage of this opportunity to get cheap land. But land reclamation was not easy. Actually, the act proved more profitable to the irrigation companies that constructed the ditches and to cattle ranchers, than it did to crop farmers. The business of both cattlemen and irrigation companies increased enormously during this period.

Twenty years after the Civil War, the western livestock industry had turned into an empire. Situated between the grain farming region (the western boundaries of Kansas and Nebraska) and the Rocky Mountains was the Great Plains. It was a region of lush short and tall grasses that had formerly nourished the immense herds of bison and pronghorn antelope. Texas beef-cattle ranchers moved their herds northward to the rich grazing lands right after the Civil War. Over the famous Chisholm Trail (which really was not a clearly marked or defined trail, but more of a direction) trampled thousands of Texas longhorn steers, heading north to the Great Plains. The cattle were fattened on the grasses and then shipped to the packing houses.

More and more cattle moved north. Soon, large cattle ranches or spreads paralleled the grain farms to the east and the mining camps to the west, and reached as far north as the Montana Territory. As long as the Great Plains remained free range, the cattlemen prospered and made large profits. When the farmers moved onto the rangelands, along with sheepherders, the situation was changed. Prairie grasses were plowed under; rangelands were enclosed with wire fences, cutting off the movements of cattle; and sheep competed with cattle for the grasses. Wars between cattlemen and farmers, and cattlemen and sheepherders mounted in intensity. Farmers, cattlemen, and sheepherders were killed. Finally, the United States Government stepped into the fight and forced the warring factions to make peace. Ultimately, the cattlemen had to accept the great change that was taking place in the West and adapt their livestock industry to the times.

Agriculture was changing in other sections of the country, as well. Improved farming practices, more labor-saving machinery, and better seeds were advancing farming in the North and South. The nation's first veterinary school at Iowa State College was training veterinarians to combat animal diseases. More agricultural information was being made available to farmers.

Largely through the efforts of Wilbur Olin Atwater, what might be called the first agricultural experiment station was established in Connecticut. The Connecticut Society for Promoting Agriculture had long before begun agricultural experimentation. Now the federal government lent its power and resources to agricultural research. Atwater was appointed director of the Connecticut experiment station, a position which he kept for fourteen years.

Agriculture progressed in all its branches. But agricultural reconstruction and expansion, although it provided employment for many people and made the transition from farm laborer to farm owner comparatively easy in the West, also created a trend toward *over*production. Farmers were now producing crops and livestock faster than they could be sold or consumed. It was a situation that was to haunt and perplex farmers for the rest of the nineteenth century and most of the next one.

7.

Into the Twentieth Century

Agriculture's progress in the last quarter of the nineteenth century was marred by some problems, most of them of an economic nature. Federal taxation, which farmers claimed brought in more revenue than the government needed for ordinary expenses, hurt agricultural people. So did state and local taxes for schools, roads, and other community needs. Many farmers felt that they were not getting a square deal. The wealth of the country seemed to be accumulating in the hands of a relatively few people, mostly eastern bankers and industrialists.

Many farmers became disenchanted with their lot. Even though the general level of agriculture was one of prosperity after 1880, the fact that vast sums of money were tied up in farmlands and railroads caused severe financial repercussions. A result was the Panic of 1884. Again, farmers were hard hit; the price of wheat dropped to sixty-four cents a bushel.

One of the most serious problems was the high cost of shipping farm products on the railroads. The continued prosperity of the Midwest depended upon the economical transportation of grain and livestock. Granger laws sought to remedy the situation, but abuses by the railroads became more flagrant. Except for some mild competition around the Great Lakes and along the Mississippi River, the major railroads enjoyed a monopoly. Consequently, they could—and did—regulate freight rates as they saw fit. Some of their abuses included lobbying for taxation favors, discrimination in freight rates so as to favor certain localities or individual shippers (by giving them lower rates), and raising rates to create dividends on watered stock. The latter abuse consisted of diluting or watering the stock in a railroad company by issuing more stock than the company's operation called for, and then raising freight rates to increase the income or dividends on the watered stock.

Farmers and the Grange fought these outrages by the powerful railroad monopolies. But the United States Supreme Court ruled in 1886 that no state had the right to enact laws regulating commerce carried on between two or more states. As a result of the ruling, the Granger laws became ineffective and were ridiculed and ignored by the railroads.

Senator S. M. Cullom of Illinois was appointed to a Senate committee assigned to investigate the troublesome interstate railroad problem. The Cullom Committee, as it was known, conducted an extensive study of the problem and recommended the creation of an interstate commerce commission to regulate the railroads. Naturally, the railroads set up a cry of protest and a battle ensued. There were many hearings and stormy debates. Finally, the Cullom Act, which created the Interstate Commerce Commission, was signed into law by President Grover Cleveland. The passage of this act eased the burden on farmers and shippers, but did not eliminate some of the abuses by the railroads. It also awakened the general public to a most important fact: The federal government could—and would, when necessary—exert its power against private industry. Moreover, the Interstate Commerce Act set a precedent for future government legislation concerning huge monopolies.

The ones who suffered most under the monopolies and offered resistance were farmers and industrial laborers. The idea of merging these two groups into one powerful force occurred to many people, including the politicians. However, all attempts to create a united front by consolidating farmers and laborers met with failure. The reason was not difficult to find: Farmers and laborers had very little in common; they were miles apart in their needs and demands.

Sharecropper and his young son working the family farm.

Farmers, particularly those in the West, wanted cheap currency (they were heavy borrowers of capital) and cheap transportation. The farmers did not work for wages; therefore, they were not concerned about wage increases or shorter hours. The nature of farming made it necessary for farmers to toil long hours in the fields and barns. Also, western farmers were vitally interested in what happened to public lands, especially when some of the best acres were being handed over to the railroads or illegally acquired under the Homestead Act. Farmers needed field hands and therefore encouraged immigration.

Laborers, on the other hand, wanted wage increases, shorter hours, and better working conditions. They had little interest in what happened to lands in the West. And labor unions wanted to keep cheap immigrant labor out of the country.

Discontent over the farm situation increased each year. Cheap land, increased immigration, improved farm machinery, and extension of the railroads caused an escalation in the population of the farming states and territories. Production kept going upward. But now the flaws of this unbridled, surging farm production became more prominent. Farm prices dropped lower and lower. Soils both in the North and South showed stronger signs of erosion and infertility. Some farmers in semiarid regions experienced crop failures. High mortgages and interest rates were hurting farmers all over the country.

Although the economics of agriculture were in serious difficulties, the scientific and educational phases of farming were making good progress. In 1889, the Department of Agriculture was promoted to the status of a Cabinet office, with appropriations for funds and a Secretary of Agriculture, the first of whom was Jeremiah M. Rusk, former Governor of Wisconsin.

Subsequent to its elevation in rank, the Department of Agriculture embarked on a variety of agricultural programs that exerted a helpful influence on farming in America. It collected statistics and other data on many areas of agriculture, such as entomology, agricultural chemistry, forestry, vegetable production, fruit culture, agrostology, livestock diseases, soils, and irrigation. A survey on the economic importance of wildlife was also undertaken by the Department, and more than five hundred publications on farming and related subjects were published and distributed to farmers and other interested persons. A *Yearbook of Agriculture*, which encompassed all areas of agriculture in the nation for the preceding year, made its appearance at this time and became an annual project.

Though few in number, agricultural scientists were hard at work searching for ways to improve crop and livestock production. Mark Alfred Carleton, a Department agronomist, devoted his time and energies to finding hardier varieties of wheat in other countries, wheat that could be grown successfully in the rugged climate of the northern sections of the wheat-producing region of the United States. Largely through the efforts of Carleton, hard and durum wheats were introduced to grain farmers in the Midwest.

Jake Leaming and James L. Reid, plant specialists, concentrated on developing new varieties of corn for maximum yields. Marion Dorset, a biochemist, made some important contributions to the improvement of livestock, the meat and dairy industries, and public health. A serum against hog cholera developed by Dorset greatly reduced losses from this disease, which cost swine producers millions of dollars every year. He was also the first scientist to make a chemical analysis of the tubercle bacillus.

Dorset became chief of the biochemic division of the Bureau of Animal Industry (BAI). In this position, he organized a system of federal inspections of pharmaceutical firms licensed by the government to manufacture serums, viruses, toxins, and related biological products. Dorset also drew up procedures for use in the administration of the Federal Meat Inspection Act.

A most important addition to the Department of Agriculture was the establishment of a veterinary division, with Dr. D. E. Salmon as director. Under the leadership of Dr. Salmon, dedicated scientists such as Cooper Curtice, F. L. Kilbourne, and Theobald Smith waged war on animal diseases, especially those communicable to man. Tick fever, hoof-and-mouth disease, pleuropneumonia, rinderpest, scabies, blackleg, and brucellosis—all came under the microscopes of the scientists. Disease-eradication programs were created and set in motion in areas showing the highest incidence of infection.

Soil specialists were rare in the Department of Agriculture, but the few who were available investigated soils of abandoned or poor farms in the North and South. More farmers were turning to the Department of Agriculture for help in restoring their worn-out or eroded soils. James Wilson, Secretary of Agriculture in the McKinley Administration, wrote in the foreword to the 1889 *Yearbook of Agriculture* that soil scientists and plant physiologists were extremely rare and that the country's agricultural colleges would have to train more of them.

Wilson, an Iowa farmer and former professor at Iowa State College, had been elected to the House of Representatives, where he served on the Agriculture Committee. Later, as Secretary of Agriculture, he extended the scope of the Department, advanced its scientific status, increased the professional staff, encouraged farm demonstrations, implemented cooperative extension work, started investigations in agricultural economics, and probed into the problems of farm credit.

When the national forests became the responsibility of the Department of Agriculture, Wilson, along with Gifford Pinchot, chief of the Forestry Service, placed forest management high on the priority list. Under Wilson's guidance and inspiration, the Departmennt of Agriculture developed into an exemplary department—an outstanding research, educational, regulatory, and custodial government agency.

Secretary Wilson worked in close cooperation with Seaman A. Knapp, a teacher, farmer, and publisher. Knapp urged southern farmers to give up the one-crop farming system which was ruining their soils and become diversified in their farming practices. Wilson sent Knapp overseas to study and obtain new varieties of rice; on his return, Knapp planted rice in Louisiana as an example of an alternative crop for the southern planters and farmers. He set up demonstration farms to show that crops other than cotton and tobacco could be profitably grown in the South. These demonstrations evolved into the Farmers' Cooperative Demonstration program, which was later expanded to include club projects for farm youths and homemaker demonstrations. Knapp was a powerful influence in extending and expanding agricultural education on a practical level.

Agricultural education at the college level was not in a slump; it was progressing slowly in important areas. At their level of enrollment and budgeting, however, the agricultural colleges could not supply the nation with much-needed scientists, or undertake soil investigations and other projects to help farmers. This situation led to a conference in Washington, D. C., which was attended by representatives from the agricultural colleges and state agricultural agencies. Out of this conference came the conclusion that federal aid was needed for agricultural investigations and experimentations. The conference leaders recommended that the federal government extend aid to the states and territories for this important purpose.

Congress responded by passing the Hatch Experiment Station Act in 1887. This act authorized the establishment of experiment stations in all of the states and territories. Funds to operate the stations were to be derived from the sale of public lands. An Office of Experiment Stations was created in the Department of Agriculture, with Wilbur Atwater of the Connecticut station as the first director. With this new arm of the Department, farmers in the states and territories could obtain technical advice and assistance on soil, plants, and other farming problems.

A movement to promote the teaching of agriculture in elementary and high schools was initiated in the late 1880's. At the time the agricultural colleges were established there were very few high schools accessible to farm children. Therefore, farmers thought that their sons who graduated from elementary school should be admitted to the agricultural colleges. Some of the land-grant colleges did admit elementary-school graduates and kept the entrance requirements low enough to permit entry. Also, two-year courses in agriculture, as well as shorter courses, were incorporated into the land-grant college curricula, thus providing agricultural education to a wider group of students.

Education in the South was a major problem on all levels. Since the South continued to practice segregation, it was necessary to set up a dual school system; that is, separate schools for blacks and whites. Segregation also extended to the colleges established under the Morrill College Land Grant Act. In those states practicing segregation, two land-grant colleges were authorized: one for blacks and one for whites.

The maintenance of two school systems was difficult for the still impoverished South. But as long as southerners insisted on separate educational programs for blacks and whites, they had to bear the additional burden. Public funds for black schools below the college level went almost entirely into the elementary schools. There were few high schools for black people. Training for teachers and other educational leaders in the South was made possible through private funds from the North.

Vocational schools were still educational stepchildren, and the few that existed in the South followed the old academic pattern of instruction. The emphasis was on liberal arts and social studies because most blacks thought that they should be taught liberal arts and social studies. This situation created problems, since most southerners believed that blacks should be taught crafts and other manual labor, not the humanities.

General Samuel Chapman Armstrong, a southerner, advocated vocational schools for black students. He founded such a school in Hampton, Virginia. Armstrong's school was sponsored by the American Missionary Association and was chartered in 1870 as an independent institution called the Hampton Normal and Industrial Institute. Its motto was Learn by Doing. The officers and teachers at the Hampton Institute were white, because General Armstrong believed that blacks needed white leadership in education.

At Hampton, academic instruction was combined with courses and practice in various trades, household tasks, and agriculture. A regular agricultural curriculum was not incorporated into the school's system until 1890.

The Virginia legislature gave Hampton Institute one third of the interest on the state's land-grant fund established by the College Land Grant Act of 1862. Hampton was not a land-grant school and was not entitled to any funds under the act. However, the Morrill Act of 1890 fixed the interest on funds received from the sale of public lands at 4 percent, which was given in equal shares to each state and territory up to a maximum of $25,000 annually. Any leftover money could be given to regular schools or those not receiving benefits from the College Land Grant Act.

Hampton Institute, therefore, was given money from the interest fund.

With the funds from the Virginia legislature and donations from private sources, Hampton Institute was able to expand its teacher-training, industrial, and agricultural instruction. Considerable attention was given to teacher training, and graduates of Hampton taught at many black schools in the South. One of the Institute's most famous graduates, Booker Taliaferro Washington, became an outstanding leader who tried to better the conditions of his black brothers by educating them.

Washington opened the Tuskegee Normal and Industrial Institute in Alabama in 1881 with an appropriation of two thousand dollars from the Alabama legislature. Tuskegee Institute was a success right from the start. It had a faculty of black teachers and a predominantly white board of trustees. Tuskegee, like Hampton, was maintained and operated mostly by private funds. The general plan of instruction was similar to that at Hampton Institute, but a greater emphasis was placed on agriculture. In spite of its agricultural courses, Tuskegee did not qualify under the College Land Grant Act for funds.

Both Hampton and Tuskegee exerted a strong influence on agricultural education in the South. Under the brilliant direction of George Washington Carver, a graduate of Iowa State College, Tuskegee Institute developed one of the world's leading agricultural experiment stations.

Carver, one of the most talented agricultural scientists in America, was born a slave on a farm in Missouri. He worked his way through Simpson College in Indianola, Iowa, then obtained a Master's degree in Agriculture from Iowa State College, where he studied under Henry Cantwell Wallace, associate professor of agriculture, and Professor James Wilson, both of whom became Secretaries of Agriculture. Wallace and Wilson recognized Carver's unusual talents and encouraged the young man in his studies and research.

George Washington Carver in his agricultural research laboratory at Tuskegee Institute.
Courtesy Tuskegee Institute

After graduating from Iowa State College, Carver served there for a short time as an instructor and researcher in agriculture. Then he was invited by Booker T. Washington to join the faculty of Tuskegee Institute as director of the Tuskegee Agricultural Research and Experiment Station.

At Tuskegee, Carver developed an outstanding research station. He also taught agriculture to black students who, at first, were not too enthusiastic about farm work. Some of them had worked hard in the fields as slaves, so why, they thought, attend school only to be put to work on a farm? Under Carver's influence and example, however, most of the students realized that there were opportunities in agriculture for black people. Some black preachers opposed Tuskegee's on-the-job-training program when they learned that students had to work on the farms. They condemned the idea as one that would keep the Negro in subservient manual labor. Instead, they advocated the teaching of the liberal arts.

Carver's research and experiment station became more famous as this black agricultural scientist discovered new uses for farm products. His list of discoveries and inventions is a long one. He developed wood stains, wallpaper, paving blocks, cordage, fiber for rope, several types of flour from soybeans, and more than 118 products derived from sweet potatoes. From the peanut, he produced 300 products, making the goober a major crop in the South.

Many black farmers visited Tuskegee to seek advice and help from Carver, who was as well versed in practical farming as in agricultural chemistry. He increased the average yield of sweet potatoes from 37 bushels per acre to 266 bushels, and wrote and published bulletins on sweet-potato culture. He also designed what became known as the Jessup Wagon, a mobile agricultural demonstration unit that traveled all over the South, instructing black farmers in plowing, planting, cultivation, and other farm practices. He devised a method of storing sweet potatoes by inactivating the enzymes.

Carver's inventions and research changed the pattern of southern agriculture. His contributions to agricultural chemistry alone helped to expand the economy of the South. He was sought out by many famous men of his time. Honors came his way; he was elected to Fellowship in the British Royal Society, awarded the Spingarn Medal for distinguished research in agricultural chemistry, given the Roosevelt Medal for Distinguished Service in the Field of Science, and made a Doctor of Science at Simpson College and the University of Rochester.

Inspired by Carver, more black students enrolled in the agricultural college, although the total enrollment was not high.

In 1903, the Office of Experiment Stations launched a survey of state colleges and schools teaching agriculture to black students.

Only those schools and colleges receiving benefits under the Morrill Act of 1890, that is, money from the interest accrued on the sale of public lands under the College Land Grant Act, were included in the survey. A primary purpose of this survey was to determine the quality of agricultural instruction and facilities in the black schools.

With the exception of Tuskegee and Hampton Institutes, the black agricultural schools were of a low caliber. The agricultural courses taught at these schools were generally of a poorer quality than the industrial arts and liberal arts courses in the same schools. There were some highly competent teachers of agriculture in the black schools, but the majority of teachers were ill-suited for the work. Too many of them were either farmers without any special educational training or aptitude for teaching, or men with just a theoretical education. Many of them were simply too scientific in their approach to teaching agriculture; a number of them could not adapt their teaching methods to the special needs of black students or the conditions existing in southern black agricultural schools.

As for educational facilities and materials, the survey showed that there was a lack of these in nearly all the black agricultural schools. Teaching aids, such as charts, pamphlets, and textbooks, were hard to come by, and those that were available were often outdated. Only Tuskegee and Hampton had what might be called adequate research facilities or laboratory equipment for teaching agriculture. And those schools that had farms with modern equipment and which practiced diversified farming were the exception rather than the rule.

Public and government interest in black agricultural schools increased after 1915, especially after George Washington Carver and his agricultural research station had achieved world fame. Agricultural education for both black and white students was given impetus by the passage of the Smith-Lever Cooperative Extension Act. This act increased the extension work (programs, projects, and services for farmers) of agricultural colleges and changed the relationship between the federal government, agricultural colleges, and rural communities—a change that resulted in better teamwork between all three.

Basically, the Smith-Lever Act provided that extension work would consist of giving instruction and practical demonstration in farming and home economics to persons not attending agricultural schools. The sum of ten thousand dollars was to be given annually to each state and territory accepting the provisions of the act. Additional sums, in proportion to a state's or territory's rural population, were available. Most important of all, the county agricultural agent system, which would have its headquarters in the agricultural colleges, was created as a means of carrying out the extension program.

The great value of the Smith-Lever Act was that it brought farm demonstrations to the farmers. Many southern farmers, blacks as well as whites, profited by the instruction and demonstration work of the agricultural extension service. Agricultural colleges also benefited from the act, since they received funds and other support from the federal government.

Four years later, the Smith-Hughes Vocational Education Act was passed by Congress. This act provided for the training of agricultural teachers and had a marked effect on agricultural education and practice. With a new supply of teachers, public and vocational schools increased their agricultural instruction. High school students were introduced to the science of agriculture. Even if these students did not plan to go on to agricultural colleges—and many of them did not—they at least had some formal training and could go back to the farms and apply what they had learned.

With the land-grant colleges, agricultural experiment stations, extension work, and agricultural courses in high schools, American farmers now had four extensive educational programs to help them improve their agricultural skills.

8.

Troubled Farmers

In the first quarter of the twentieth century, agriculturists and conservationists turned their attentions to the state of the land resources. In an incredibly short time, the federal government had disposed of a large portion of its uninhabited lands to settlers, railroads, lumber companies, and mining interests. Vast quantities of the country's natural resources—soil, water, forests, and wildlife—had been squandered in the agricultural and industrial growth of the United States. Soils in certain sections were already worn out or on the way toward infertility because of exhaustive farming practices or neglect. The cut-and-move-on system of the logging companies had denuded large tracts of forest lands. Wildlife suffered a severe loss of habitat in many regions. And little had been done in the way of conserving water, which, in combination with soil, was the lifeblood of agriculture.

President Theodore Roosevelt's first message to Congress after he assumed office contained a section on the state of the country's natural resources. He spoke about the need to conserve forests, minerals, soil, water, and wildlife, stating that conservation was "the most vital internal question of the United States." Roosevelt and Gifford Pinchot, chief of the Forest Service, formed a conservation team that did much to halt, or at least slow down, the waste of natural resources.

At the turn of the century, Congress had passed the Carey Act, which authorized states in arid regions to take a million acres and place them in the hands of private companies for improvement by irrigation. Some arid lands had been reclaimed through irrigation, but in general this program had not worked out too well. Considerable money was needed for dams, reservoirs, and ditches; little of it had materialized, and when Roosevelt came into office, the irrigation program lagged far behind the goals set.

Roosevelt obtained the passage of the Reclamation Act of 1902. This law authorized the sale of public lands in sixteen mining and grazing states and territories. Proceeds from the sale were to go into a special irrigation fund. The irrigated lands would be sold to settlers at reasonable prices, on an installment plan over a ten-year period. The payments from the settlers would be used to renew the irrigation fund. Under this plan, large tracts of land, once worth only a few cents an acre for cattle grazing, increased in value to as much as several hundred dollars an acre as a result of the irrigation program.

With the continued development of farm machinery and equipment, farming became easier and more efficient in all regions. There were now horse-drawn potato diggers, disk harrows, vacuum-type milking machines, potato planters, corn pickers, vegetable transplanting machines, and crude steam-engine tractors. As American technology increased, much of the hard labor and drudgery was being taken out of farming. The mechanical revolution raised the productivity of farmers to a point where more and more farm products were being exported.

World War One further stimulated production, to a degree where farmers produced more crops and livestock than ever. The federal government encouraged increased production, telling the farmers that overproduction was nothing to fear, for America would be feeding a hungry Europe for a long time. A slogan was made up: Food will win the war—don't waste it! Under this slogan and with government encouragement, America's farmers and ranchers doubled their efforts to bring forth great harvests. Many farmers bought new lands to place under cultivation to meet the increased war needs for food and fiber. They hired as many workers as they could find, although industry and the military services siphoned off most of the younger men.

The scarcity of farm labor during World War One did two things: (1) it speeded up the development of more labor-saving farm machinery, and (2) it emphasized the importance of *skilled* farm workers. The value of a farm worker in 1914–1917 was determined by his skill and reliability, not by his physical strength. Farm machines and draft animals could provide the brute strength for the heavy work on the farm. What farmers needed were workers who could do the various farm jobs with speed and efficiency.

America's farmers did produce enough food to feed millions of people in war-torn Europe. When the war was over, farmers found themselves with surplus crops and no foreign markets. People in Europe were still hungry and many of them were on starvation rations. But most of the countries were impoverished; they lacked the money to buy American farm products. Also, a high American tariff made it difficult for those countries which did have some money to buy American food and fiber. The tariff was helpful to manufacturers because it held their prices up. This was not so in the case of farmers; it lowered their prices. At the same time, it forced farmers to purchase their supplies at tariff-protected prices. Thus, the familiar problems of farmers—low prices for farm products and high costs of supplies—returned to haunt them after World War One.

Even though farmers had doubled their output during and right after World War One, this did not mean that every farmer produced a bumper crop with efficient and improved farming practices. There were good and bad farmers. Furthermore, sharecroppers simply did not have the means to produce crops on the same level as the large or more mechanized farmers.

The small one-mule farmer was in a predicament. He managed to plow perhaps ten or twelve acres in as many days. If he wanted to get ahead of the game and plowed his fields in the fall, winter rains and spring thaws washed away the shallow soil or packed it down hard, causing extra work later on.

High-priced or improved seeds were beyond the means of

the poor farmer or sharecropper. He bought ordinary corn, cotton, or other seeds, and a minimum of fertilizer with money obtained by mortgaging his future crop. Cultivation of his crop was laborious and time-consuming, requiring backbreaking work in the fields from sunup to sundown. The sharecropper and his family were kept busy walking up and down the crop rows cutting down weeds. There was little or no time for other important work, such as correcting harmful practices that destroyed soil fertility or checking soil erosion.

One-mule farmers obtained only minimal crop yields, perhaps only one third of a bale of cotton or ten bushels of corn per acre. And they produced those low yields at a high cost to themselves and their lands. Returns on crops barely paid for the rent, seed, fertilizer, and food and clothing for the sharecropper family. And the fields kept getting poorer and poorer.

There were many of these one-mule farmers in the South and Southwest. Part of their problem was caused by circumstances, part by ignorance and a reluctance to change or improve their farming practices. Regardless of the causes, the poorer farmers were becoming a major problem.

In contrast, the larger and more efficient farmers took advantage of all the advances in agriculture. They sought technical assistance, read the latest farm bulletins and pamphlets, and used new and improved farm machinery. As a result, their crop yields more than justified their efforts.

But these progressive farmers were in the minority. Therefore, a growing concern was the lack of efficient farm management. The agricultural colleges, experiment stations, and the Department of Agriculture doubled their efforts to research and recommend good farm-management practices. The thrust of their programs was to show farmers how to make farming more profitable and at the same time to conserve soils and build up fertility.

Soil depletion, particularly in the South and Southwest, was now a serious problem. The single-crop system of the South had been extremely hard on the soils. A lack of adequate drainage of soils along the Atlantic coast was another factor that was preventing potentially productive soils from being placed under cultivation in the South.

In the Southwest, valuable soil was lost to erosion caused by high winds and heavy floods. Each year, farmers were faced with the problem of how to stop the loss of their soils. A similar problem existed in the Midwest, although soil erosion had not reached the proportions found in the Southwest.

There was a variety of soils in the Northeast, all capable of growing diversified crops. But northeastern farmers were also experiencing soil problems. Here, the task was to utilize soils so that they produced crops that yielded large returns per acre, since there was not as much farmland as in the Midwest. Many northeastern farmers adhered to the old practice of growing grain and hay. Hay production, of course, was beneficial to soils, provided that the fields were kept in grass. A large number of northeastern farmers plowed under their grasses and planted grains. The grain crops extracted a great deal of fertility from the soils and laid them open to erosion.

A national soil survey, begun near the end of the nineteenth century, was far from complete in 1920. The analysis and recording of every soil type in the United States was a gigantic task, a project that was carried over from year to year. As the soil survey progressed, one fact stood out: In many areas, the nation's soil was in trouble from inefficient or harmful farming practices and the forces of erosion.

The economic plight of the average farmer grew steadily worse. He was still caught in the familiar cost-price squeeze, with the price of wheat, corn, and hogs declining year after year. The actual cost of growing a bushel of oats in 1922 was more than the sale price. And the purchasing price of the farmer's dollar had shrunk to less than seventy cents.

Unable to extricate themselves from this financial fix, the farmers called on the federal government for assistance. One hundred representatives and senators of both political parties from the major agricultural states formed a farm bloc. This group of agriculture-minded legislators pressed Congress for farm relief. Their influence and efforts resulted in the passage of a number of measures: laws to prevent speculation in future grain supplies; protection for dairy farmers against adulterated milk; legalization of farm cooperatives; and fair treatment for livestock farmers by the agents of stockyards and packinghouses.

Congress appropriated over $430 million to finance the export of farm products. The passage of the Agricultural Credits Act made the credit of the Federal Land Bank available to farmers who wished to borrow on the security of their livestock and crops being readied for market.

The demands of farmers produced the McNary-Haugen Bill, which proposed a National Agricultural Export Commission. This commission would be authorized to buy up surplus farm products and sell them abroad for whatever they would bring on the market.

Charles Linza McNary, the son of an Oregon pioneer farmer, was deeply concerned with the welfare of American farmers. He advocated and worked for many important agricultural and conservation programs, among them reforestation, irrigation, land reclamation, and development of waterpower. He introduced the McNary-Haugen Bill into Congress.

Considerable controversy flared up over the McNary-Haugen Bill. Opponents claimed that the government could not legally help the farmers with supportive legislation. The fact that the railroads and other industries had been so helped by the federal government was forgotten or ignored by those who opposed the bill. As a result of the opposition, President Calvin Coolidge vetoed the bill.

There was a general prosperity during Coolidge's Administration, with one notable exception: agriculture. As a group, America's farmers had grown poorer while the rest of the nation prospered. The economic plight of agriculture had become so bad that farm people were leaving the farms, moving to the cities, and finding jobs in other industries. In the five years between 1922 and 1927, more than a million people left the farms. Farming in the late 1920's was an unprofitable occupation, for the average net income of the farmer was only nine hundred dollars a year.

Congress tried to ease the farm crisis by passing a second McNary-Haugen Bill in the spring of 1928. The new or revised bill contained most of the features of the original and it, too, met with stiff opposition. Opponents claimed that the bill would put the government in the farming business, since the government would purchase surplus cotton, tobacco, grain, and livestock.

The bill also had what was called an equalization fee. In the case of wheat, the equalization fee was in the form of a five and a half cent tax on a bushel. Its purpose was to help reduce the loss

These cartoons illustrate the plight of the farmer during the 1930's.
Rollin Kirby in Scripps-Howard Newspapers

Some of the farmer's regular boarders.

Reprinted, courtesy of the Chicago Tribune

that the government would experience by buying wheat at one price and selling it for less. This equalization fee or tax would reduce the net sale price to farmers, but they would still be better off than without any sales or sales at lower prices.

After much debate and deliberation, President Coolidge again vetoed the bill. His action caused a great deal of resentment among farmers, especially those in the grain-producing region. They thought the President of the United States should show more understanding for their difficulties.

When Herbert C. Hoover assumed the presidency in 1929, the economic condition of agriculture had worsened; furthermore, the financial problem had become more widespread.

During the presidential campaign, both the Democrats and Republicans had promised the farmers relief from their financial troubles. A short time after taking office, Hoover moved to keep his party's pledge to the stricken farmers, approving an Agricultural Marketing Act as passed by Congress and signing it into law.

This act created a Federal Farm Board of nine members, including the Secretary of Agriculture. It provided a revolving fund of $500 million from which loans—at a maximum interest rate of 4 percent—could be given to farm cooperative associations. The loans were to be used to market cotton, grain, livestock, dairy products, and fruits to the best advantage. The Federal Farm Board also proposed to encourage more scientific farm management on individual farms, more or less along the lines of operation by industries.

At first, the Agricultural Marketing Act seemed to be the answer to the farmer's economic dilemma, or at least to his marketing problem. Loans in the amount of $165 million were made to 132 farmer cooperatives. But prices for farm products kept going down, and borrowing money did not help; it only served to increase the farmer's burden.

There was no provision in the Agricultural Marketing Act which authorized the federal government to buy and sell surplus farm products. Nevertheless, the government did take large surpluses of cotton and wheat from farmers in an attempt to stabilize prices. At the end of 1933, the government had spent about $375 million on wheat alone, which had fallen in price from a dollar to sixty-one cents a bushel. Cotton prices also declined to a new low, from sixteen cents a pound to six cents. Consequently, the government ended up with a loss of about $185 million and was severely criticized by those who opposed federal aid to farmers.

Prices kept dropping, and the farmers' situation became worse. Farm leaders in Washington pleaded the farmers' cause with eloquence, fervor, desperation, protest, and exhortation. Their efforts were to no avail. Congress was unable to come up with any remedial programs. But now the economy of the entire country was in serious straits, for the stock market crash of 1929 had triggered a general depression.

The complaints and mutterings of farmers grew louder. An Alabama planter named Ed O'Neal was elected president of the Farm Bureau, a national farmers' organization. O'Neal, a tall, white-haired man, supplied vigorous leadership to this group of farmers. Another farm organization, the National Farmers' Union, appointed a new leader, John A. Simpson, who had a violent hatred of international bankers. Simpson blamed the bankers for the desperate economic plight of American farmers. Both O'Neal and Simpson carried the farmer's cause to the smoke-filled rooms of congressional committees and demanded immediate action.

But the pressures of the Farm Bureau and the National Farmers' Union brought no action from an administration which thought there was no solution to the farm problem. Many officials in the Hoover Administration believed that the troubles of the farmers, like those of poor people in the cities, would always be present, no matter what was done.

Some Iowa farmers thought otherwise and took matters into their own hands. Led by Milo Reno, a former president of the Iowa Farmers' Union, a number of farmers created the Farmer's Holiday Association. The idea behind this organization was that farmers were entitled to a period of time during which they could straighten out their financial affairs free from the responsibility of making payments on debts. Banks and businesses had received this kind of holiday after the stock market crash; therefore farmers should be given the same opportunity, according to Reno and other Iowa farmers.

Reno, a big man in his sixties, was slow-spoken and friendly, but he was a fighter. He had been on the agricultural battlefield for a long time. Reno believed that if the farmers could not obtain justice through fair legislation, they had the right to refuse to deliver farm products at less than production costs. What he meant, of course, was that farmers had the right to strike for better prices.

Under Reno's dynamic leadership, some Iowa farmers did go on strike. They blockaded roads near Sioux City, stopping trucks loaded with farm produce. The striking farmers convinced those in the trucks that it was to their best interests not to market their crops and to turn around and go back to their farms. Those farmers who did not want to turn back were forcibly stopped, or their truckloads dumped onto the roads and destroyed.

Dairy farmers joined the strike. Milk distributors in one Iowa county had been buying milk from farmers at two cents a quart and selling it for eight cents. At this price, it was hardly worth the dairy farmer's efforts to produce milk, so they went on strike for a better price. They placed an embargo on all milk entering Sioux City, except for deliveries to hospitals. Soon the highways into Sioux City were patrolled by striking dairy farmers, who—unlike the vegetable and fruit farmers—pursued strong-arm tactics, opening milk cans and dumping the contents on the road. If necessary, the striking dairy farmers roughed up any farmer who did not go along with the strike. The milk strike spread into other Iowa counties and eventually into other states. Eastern dairy farmers joined the strike, and rural roads in Pennsylvania, New York, and New Jersey ran white with dumped milk.

The farmers' strike of 1932 did not bring any relief as far as raising prices was concerned. Although milk was in scarce supply and prices rose, the release of more milk when the strike was over soon drove prices down again. But the strike did have some benefits for the beleagured farmers. For one thing, it gave them a chance to get rid of their bottled-up anger and frustration, if only for a short time. And, on a larger scale, the farmers' strike pointed out the paradox of a great agricultural and industrial nation that was unable to make intelligent use of the food and fiber it produced. People were going hungry because nobody had been able to figure out a way to balance farm production with the needs of the nation.

9.

Agriculture's New Deal

The American farmer, like his counterpart in other countries, followed his profession with a set of fears. He dreaded floods, drought, insects, diseases, killing frosts, crop failure, low prices, unfair legislation, and high transportation rates. Now—in 1933—two new fears were added: *overproduction* and *soil destruction*.

Several centuries of intense and unwise farming practices had caught up with the land in many sections of the country. Out of the 1,903 million acres which comprised the continental United States, about 413 million were under cultivation. Approximately one half of these acres were in farmlands; slightly more than one fifth in forests; and a little less than a fifth in grasslands. But now nearly 35 million acres of former cropland were unfit for farming. About 450,000 farms, representing 75 million acres, were incapable of supporting the families that worked them. These worn-out farmlands were a stark reminder of the wasteful practices that had brought American agriculture to its high productive peak.

Loss of fertility was now common in northern and southern soils, and erosion by wind and water compounded the soil problem. The Midwest and West still had relatively fertile soils, but the former prairie lands now showed scars of intense cultivation with heavy machinery. Erosion, given a foothold by machine farming and activated by wind and water, fingered its way over the eastern prairies. Exposed to floods and high winds, the prairie soil, no longer securely anchored by tough grasses, was washed or whisked away. Valuable topsoil, hurried along by floods, flowed into creeks and rivers and mixed with the sterile mud. Then it traveled away from the prairie region to build up at the confluences of rivers or to form deltas where the Mississippi met the Gulf of Mexico.

Each year, the prairie soils felt the bite of the plow and the scratch of the harrow; they were gouged, pulverized, and broken down and left to the caprices of wind and water. "Bare ground left by the plow will have as much soil washed off in ten years as the unbroken prairie will lose in four thousand," wrote Paul B. Sears in his somber book *Deserts on the March*, published in 1935. This book told Americans, or at least those who bothered to read it, what was happening to America's soils in the 1930's.

Wind erosion on the prairies was aggravated by the increased use of power machinery, which broke down soil structure. Out on the rangelands, overgrazing by cattle and sheep caused conditions leading to severe erosion. Overgrazing and other harmful practices of ranchers were creating insidious, adverse changes in the ecology of the prairies. One of the practices was the repeated burning off of dead grasses. It radically altered the composition of the prairie soils. The changes made by overgrazing and burning reduced or eliminated yields of grasses in many sections and affected the quality of those grasses that remained.

Worn out, altered, or eroded prairie lands produced low-grade plants. Nature hastily grew poverty grass and other weeds in an attempt to cover the denuded sections of the prairies. Poverty grass, a pale-gray grasslike plant, quickly moved into bare spots after wind and water had carried off the topsoil.

Some areas of the rangelands were so stripped of the natural grasses that they could no longer support livestock and wildlife populations. Attempts had been made to reestablish prairie grasses, but this was a slow process and the results were uncertain at that. At least thirty to forty years were required to restore grasses to the prairies. Also, wind and floods impeded the grass-restoration programs, for seeded areas were deprived of soil by wind and water before the grasses could take hold.

Thus, by the early 1930's, severe soil erosion and depletion were the accumulative results of unwise farming practices. Now nature added to the troubles: Several years of sustained drought and high winds dried out the bare soils and deepened the scars of erosion. Dust storms became more frequent in Kansas, Missouri, and Oklahoma. As the drought continued, many farms lost topsoil and turned into barren places, with sand and subsoil heaped high. Fences and farm machinery were buried under earth drifts, houses and farm buildings were piled to the eaves with the displaced soil; deeper and deeper grew the dust hills as the drought and high winds devastated the lands in the so-called Dust Bowl.

City dwellers, as well as farmers in the Midwest, were affected by the dust storms, for the winds transported soil particles for long distances. New Yorkers, for instance, were amazed to learn

High winds blew the soil from farms in the Midwest and Southwest, covering machinery, fences, and buildings. *USDA Soil Conservation Service*

that the dust clouds over their city contained soil from Kansas farms. Conditions were desperate in the states located in the center of the great dust storms. The dust was everywhere. Children wore masks to protect their faces and lungs as they walked to school. Housewives battled the ubiquitous dust that blew or seeped into their houses. Respiratory ailments, aggravated by the dust, greatly increased. Visibility on any given day might be a quarter of a mile, a few blocks, a hundred feet, or less.

Midwesterners were not strangers to dust. The flat prairies were always dusty on dry days or during droughts. But the farmers had never lived through anything quite like the great dust storms of the 1930's, which whirled the soil away as easily as one blows dust from a tabletop. Many Dust Bowl farmers were ruined and simply packed up and abandoned their destroyed farms.

Franklin Delano Roosevelt's Administration inherited the serious farm problems, as well as the depressed economy of the country. Roosevelt appointed Henry Agard Wallace as his Secretary of Agriculture. Wallace was the son of Henry C. Wallace, former teacher of George Washington Carver and Secretary of Agriculture under Presidents Harding and Coolidge. The new Secretary of Agriculture had an Iowa farm background, with a Bachelor's degree from Iowa State College. He had been editor of *Wallaces' Farmer*, an influential midwestern farm newspaper, and was regarded as an authority on farming methods and problems.

Wallace was well acquainted with the needs of farmers. A little more than 5.5 million farm families were on the nation's farms. Their average income was less than nine hundred dollars a year. The bottom had all but dropped out of the farm-product pail. Wheat fell to thirty-eight cents a bushel, corn to thirty-two cents, and a pound of cotton brought only eleven cents. The total drop in value of the country's farm products was from $12 billion to $5 billion at the time when Henry Wallace became Secretary of Agriculture.

Forty-three percent of the nation's farmers were tenants

or sharecroppers. Thousands of farmers who owned their farms could not meet their mortgage payments. A large number of them lost their farms and went to work as farm laborers. Other families were evicted and moved to the cities or wherever else they could find shelter, taking with them only what they could load into a farm wagon. It was a dark time for farmers and their families all over the country, and no one could see any light ahead.

The situation of the sharecropper was a dismal and desperate one in 1933. There were about 1,100,000 white and 700,000 black sharecroppers in the South. Sharecroppers could look forward to nothing but poverty and drudgery. A black sharecropper worked an average of ten or twelve acres; his patch was just one of many on a large cotton or tobacco plantation owned by landlords. The farmland and house were rented to the farmer. He was loaned a mule and plow, and given cotten seed or tobacco seed and fertilizer. All the work in producing the crop, from seedbed to harvesting, was performed by the sharecropper and his family.

When harvested, the crop was delivered to the landlord, who sold it and collected the money. The sharecropper and landlord were supposed to share the proceeds on a fifty-fifty basis. But the sharecropper rarely saw any money. He was in debt up to his ears and had to pay interest on everything he had borrowed. Grocery bills and clothing for his family ate up whatever money was left after the loans were paid. Usually the sharecropper's food and sundries were obtained on credit. If a sharecropper was lucky, he might net sixty dollars for a year's work; if luck ran away from him, he ended up in the hole, owing money on next year's crop before it was planted.

Southern tenant farms were the rural ghettos of America. Sharecropper families lived in ramshackle, unpainted clapboard houses, with just the bare essentials of civilization: decrepit furniture and a few pots and pans. Corncobs were burned as fuel to keep the families warm. Cornmeal, molasses, and fat pork made up the daily ration of the sharecroppers. White sharecroppers were not much better off than their black counterparts. It was a time

Most sharecroppers and tenant farmers lived in unpainted, decrepit houses, such as this one in Louisiana. *USDA Photo*

of want in a land of plenty. It was a time when, as Franklin Roosevelt declaimed in one of his fireside chats heard over millions of radios, one third of the nation was ill-clothed, ill-housed, and ill-fed.

Out in the Midwest, mortgage holders were foreclosing on farmers who were behind in the payments. Eight hundred Plymouth County, Iowa, farmers staged an open rebellion against foreclosure. The desperate farmers banded together and physically prevented a sheriff and judge from foreclosing and taking possession of their farms. This action of the embattled farmers served to awaken the country to the almost hopeless condition of rural America.

Secretary Wallace believed that the only way for the farmer to maintain his income in a free market was to reduce crop yields at strategic times. Overproduction had brought the farmer to his present financial crisis. In the days when Wallace was editing

Wallace's Farmer, he had called for government aid to farmers. He had recommended government crop insurance and urged voluntary crop reduction by farmers. As early as 1922, he had proposed what he called the ever-normal-granary plan for the storage of surplus farm products.

Wallace and other agricultural economists suggested crop-control measures that were to become part of the Agricultural Adjustment Act passed by Congress in 1933. Under this act, farmers would receive allotments or quotas for production and would be paid benefits for acres not planted in cash crops.

Henry Wallace's ever-normal granary was adopted. It was patterned after a plan used by the biblical Joseph, who was placed in charge of the Pharaoh's granaries. Joseph worked out a system whereby grain was stored in good crop years, and removed and sold in bad years. Wallace's counterpart of Joseph's granary plan would see to it that there was the same supply of grain and cotton on hand and for sale year after year.

There was opposition to the ever-normal granary. Too many good crop years would cause the granaries to overflow, claimed opponents. Crops would be wasted. Also, the federal government would have too much power and control over farmers. And Wallace's plan would cost the taxpayers money.

Debates over these bold farm-relief programs became stormy. As a result of the heavy opposition, the Agricultural Adjustment Act was declared unconstitutional by the United States Supreme Court. In the opinion handed down by the Court, it was stated that the processing tax (which provided the money to pay benefits to farmers) was an unconstitutional use of the taxing power of the federal government.

Congress quickly followed up with the Soil Conservation and Domestic Allotment Act. To avoid the unconstitutional tax problem of the AAA, the drafters of the new measure called for payments to farmers who put a portion of their crop-producing acres into grasses, legumes, and other soil-building or conservation crops. The new law provided price supports and soil-conservation measures, two supposed cures in one medicine for major farm problems.

The Soil Erosion Service was created in the Department of the Interior under the supervision of Dr. Hugh H. Bennett, known as Big Hugh. He directed soil-erosion investigations, which were carried out with the cooperation of agricultural colleges and experiment stations. More than forty soil- and water-conservation demonstrations, using fifty Civilian Conservation Corps units, went into operation during the first eighteen months of the Soil Erosion Service's existence.

In March of 1935, the Soil Erosion Service was transferred to the Department of Agriculture and renamed the Soil Conservation Service. The organization and scope of the service was expanded to include all soil and water conservation activities. Bennett and his staff set in motion an integrated war on soil erosion and depletion. Their slogan was: Use each parcel of land according to its capability and treat it according to its needs.

The Soil Conservation Service offered technical assistance to farmers, creating soil-conservation districts in all of the states, with soil and water specialists assigned to each district. It assumed the leadership in the soil survey begun in the nineteenth century. On-the-farm drainage and irrigation demonstrations were conducted by Soil Conservation Service specialists. The service supervised the water utilization program in western states, including coordinated snow surveys for forecasting water levels and resources, both of which were important to western farmers and ranchers.

More Americans were becoming aware of the extent of soil erosion through the work of the Soil Conservation Service. In a Soil Conservation Service pamphlet entitled *To Hold This Land*, Russell Lord stated that Americans had "manhandled the land." He explained that although undisturbed soils have some erosion, it is not too serious, and nature attempts to correct the condition. But when the natural vegetation of soils is removed, when trees are cut down, prairies plowed up, when soil is pulverized with steel instruments and deprived of native spongy organic growths that make it absorptive, then the natural soil balance is destroyed, and flash floods, high winds, and other erosion forces remove soil faster than nature can replace it.

While AAA programs tried to cope with low prices and overproduction and the Soil Conservation Service worked to conserve and restore the nation's eroded and depleted soils, the Resettlement Administration attempted to rehabilitate distressed and displaced farm families. Foreclosed mortgages and dust storms had made many families homeless and destitute. Opposition to the resettlement program hampered the work of the agency and the results were unsatisfactory.

The tenant-farmer problem was increasing and the government wanted to reverse this trend, since it was contributing to poverty in the South. John Hollis Bankhead, Senator from Alabama, cosponsored the Bankhead-Jones Farm Tenant Act of 1937, which established the Farm Security Administration. Loans were offered to farm tenants to make them change from tenant to farm owner. However, too many of the tenant farmers and sharecroppers were overloaded with loans and other debts. They could or would not take on any more financial burdens. This program, like other New Deal farm programs, met with great opposition, especially from southern white farmers and planters who objected to such aid for blacks. Eventually, the Farm Security program failed for lack of support.

Another important New Deal program for farmers was the work of the Rural Electrification Administration. City and town dwellers enjoyed electricity and its by-products. But most of the nation's farmers were still in the kerosene-lamp and lantern age. Electric power lines bypassed farmers because the costs of bringing power lines to farmers interfered with profits.

George William Norris, Senator from Nebraska, was the champion of rural electric power. Norris was always active in improving the conditions of America's farmers; he worked for progressive farm legislation, advocated farm relief, called for the conservation of natural resources, and believed in public ownership of generation, transmission, and distribution of hydroelectric power.

Norris sponsored the Tennessee Valley Authority Act of 1933, which established a hydroelectric power plant entirely owned by the federal government. The purpose of the TVA, as the

Authority was called, to develop the economic resources of the Tennessee River valley, to conserve its soil and water resources, and to make cheap electric power available to farmers and other people in the region.

In addition to providing hydroelectric power, the TVA also aided other farmers in the country: two munitions plants at Muscle Shoals, Alabama, were converted into fertilizer experiment stations. Fertilizers produced at the Muscle Shoals site were used for demonstration purposes to show farmers the value of using fertilizer on croplands. They were also made available to private industries for manufacture. The fertilizer program was instrumental in introducing new fertilizers, lowering their cost, and improving manufacturing processes.

Despite the depression throughout the country, agriculture was making some noteworthy advances in the areas of mechanization, crop culture, and livestock production.

Power for farm machines and equipment was shifting from horses and mules to tractors in all farming regions. Tractors had been in use since the beginning of the century, mostly in the Midwest. But early tractors were cumbersome steam-operated machines. Tractors in the middle 1930's had evolved into compact, powerful, all-purpose machines driven by an internal combustion engine. Their use was responsible for increased production and a general improvement in farm efficiency.

Various types of tractors were developed to meet the needs of farming in different parts of the country. A popular tractor was the standard four-wheel machine adapted for use on land that could travel easily with wheel tractors, rather than caterpillar types. These four-wheel tractors were equipped with belt drives and specially engineered drawbars. (The drawbar is that part of the tractor to which plows and other equipment are hitched.)

Fruit growers were also using tractors in their orchard operations. The orchard tractor was similar to the standard four-wheel type, but was lower to permit operation closer to and under fruit trees and vines. A special shield or guard on the tractor shunted branches and fruit over and around the sharp edges of the tractor to avoid injury to fruit and trees.

Corn, cotton, and vegetable crops were now being planted and cultivated with machines pulled by a general-purpose tractor. This tractor was designed for use in row-crop farming, with wheels spaced to straddle a crop row. It was a popular tractor, and more machines were being developed for use with this tractor.

Crawlers or track-laying tractors, although not new, were being improved for heavy work on hilly land, wet areas, and sandy soils. The long tracks distributed the tractor weight over flat track shoes, thus providing good traction. Crawlers had a low center of gravity, which made them safer to use on hills or mountainous terrain. These caterpillar tractors were becoming the workhorses in preparing lands for irrigation. Semiarid soil was loose and had to be leveled before being irrigated. Farmers were also using crawlers for moving soil, for building ponds, and for soil-conservation work.

Some new machines and farm equipment made their appearance in the 1930's. Portable-sprinkler irrigation units, field pickup balers, hydraulic lifts for tractors, low-pressure tires for tractors, forage harvesters, automatic barn cleaners, and self-propelled combines were helping to speed up or improve farming practices and general efficiency.

Improved varieties of crops were also increasing the farmer's productivity. A major advance in crop culture was the development of hybrid corn. Prior to the 1930's, all corn was open- or self-pollinated. The plants in a field just pollinated themselves, and there was no control of pollen to and from any plants. George H. Shull, a pioneer in plant breeding, had discovered earlier in the century that open pollination caused corn to lose its vigor with successive generations of self-pollination. He found that vigor could be restored by crossing inbred lines of corn plants. In short, controlled pollination would produce better corn. This was an important breakthrough in corn culture, a completely new method of breeding corn. It paved the way for the development of hybrid corn.

Various plant scientists and geneticists worked on hybrid corn, which greatly increased yields. Briefly, the first generation of the crossbreeding (which produces hybrids) of inbred plants—like most first-generation crosses—turned out to be as uniform as the parent plants. But the seeds of the plants resulting from the crosses could not produce plants the second year; they produced only on the first planting. Therefore the cross had to be created *new* each year from inbred stock to maintain vigor and uniformity. Corn farmers could get more uniform ears and higher yields, but they would have to plant new hybrid seed every year; they could not save seed to plant.

Hybrid corn became commercially practical after D. F. Jones of the Department of Agriculture suggested a double crossbreeding system. Corn Belt farmers were slow to adopt the use of hybrid corn; it was a new idea and most of them did not understand the genetic principles involved. Henry Wallace developed several strains of hybrid corn, among them a variety known as Copper Cross. Gradually, the value of hybrid corn was recognized by farmers and it was planted on a large scale in the Corn Belt and elsewhere around the country.

Livestock farmers were also making advances with improved breeds of animals and more efficient practices. Selective breeding of dairy cows, coupled with more nutritive feeds and milking techniques, increased milk production. Beef, swine, and sheep producers were upgrading their stock by selective breeding practices. The fabulous King Ranch in Texas was experimenting with some new breeds of beef cattle—notably the American Brahman and Santa Gertruda strains—that promised to be more resistant to heat and ticks than other breeds. Poultry farmers were increasing egg and meat production with improved breeds of chickens and better poultry feeds.

Much of the progress of agriculture in the late 1930's was due to programs instituted by the Department of Agriculture. Never before had any government initiated such bold, far-reaching agricultural reforms and programs as was done by the New Deal administration. Basically, New Deal agricultural aid was composed

Everyone works on farms, as shown by this picture of a little girl pumping water in 1942, when few farms yet had running water. *USDA Photo*

of emergency measures geared to rescue farmers from financial difficulties and conserve the nation's soil and water resources. But many programs and projects served to advance the scientific and technological levels of American agriculture. Many mistakes were made, of course, and a number of programs were vigorously opposed and met with failure. Nevertheless, the majority of programs moved American agriculture into the role of world provider of food and fiber.

10.

Agriculture after World War Two

In 1941, the pattern of agricultural production was about to repeat itself. When World War Two involved the United States in armed conflict, American farmers were again called upon to increase their production of food and fiber. Almost overnight, agriculture changed from an ailing industry bolstered by New Deal legislation and emergency programs into a gigantic, all-out, self-sustaining effort to meet the need of a great nation at war.

To speed up production, crop controls and other restrictions were eased or set aside. Instead of receiving payment for keeping acreage out of production, farmers were now paid incentives to increase production. Sharp rises in prices of farm products were allowed, so as to stimulate production further. America became the breadbasket of the world, with the responsibility of feeding the free countries as well as arming them.

Domestic and foreign demands for food and fiber started American farmers on the road to prosperity. They soon forgot the hard times of the 1930's. Gone were the desolate, desperate days of the Dust Bowl—long droughts, displaced farm families, eroded soils, foreclosed mortgages, surpluses, low prices, and breadlines. Farming was once again on the upswing in all its branches.

One New Deal program proved its worth in the early days of World War Two. Henry Wallace's ever-normal granary provided a headstart for America's task of feeding half the world. When Congress declared war on Japan and Germany, the farm storehouses held 400 million bushels of wheat and vast quantities of corn and cotton. These commodities had been stored for release at some future date, when the total yields of these crops would be low enough for their withdrawal. World War Two created an immediate demand for the products of the ever-normal-granary.

Other New Deal programs supplied a valuable base for wartime production. Soil conservation was one of them. Although conservation programs initiated by the Soil Conservation Service had only made a dent in the over-all soil problem, they laid the foundation for future controls. The strip-, contour-, and terrace-farming demonstrations presented in the 1930's had convinced farmers—at least those who were interested in saving their lands—that sustained high yields could only be obtained on fertile and noneroded soils.

More fertilizers were used in producing the bumper war crops. Thousands of tons of lime, available to farmers under the soil conservation program, were applied to soils. Green manuring, a soil-building practice promoted by the Soil Conservation Service, was another measure employed to increase soil fertility and structure. It consisted of plowing under crops such as clover and ryegrass, which would add to soil fertility.

Increased production meant an increase in the use of heavy soil-pulverizing machinery. Therefore it was necessary to keep watch on the condition of soils. But it was easy to forget about soil conservation in the middle of a concerted effort to produce high yields. Farmers, like other users of natural resources, had a tendency to overlook the needs of tomorrow in the fever of producing today. The Soil Conservation Service acted as the watchdog of the nation's soil and water resources, helping farmers to meet their crop quotas but at the same time conserving natural resources.

Despite the mistakes and warnings of the past, wartime production was not carried out with efficiency by all farmers. There still was a core of farmers who could not or would not take steps to protect their soils. Some conservation agencies and worried individuals voiced their concern over the deterioration of the soils. Edward Faulkner, an Ohio county agricultural agent, wrote a startling book in which he indicated the plow as the main cause of most soil-erosion problems. Faulkner's book, *Plowman's Folly*, presented clear, logical evidence that the moldboard plow was the least satisfactory implement for preparing lands for crop production.

> Had we not gone contrary to the laws of nature by plowing the land [wrote Faulkner], we would have avoided the problems as well as expensive time-consuming efforts to solve them. That we would have missed all of the erosion, the sour soils, the mounting floods, the lowering water tables, the vanishing wildlife, the compact and impervious soil surfaces is scarcely an incidental consideration.

Faulkner's arguments went on to suggest that the moldboard plow be discarded as the initial soil-preparation implement. He advocated plowless farming by using disks that would cut up crop wastes or green manure and work them into the soil. By doing so, farmers would not only add to the structure of their soils and reduce erosion, but would restore the vital capillary action of water and plant nutrients, which was disturbed or broken down by plowing.

This was a thought-provoking and custom-shattering idea. For thousands of years, farmers had plowed their lands, despite the fact that nobody had ever come forward with a scientific reason for plowing. It was a tillage practice that had been handed down from generation to generation. Consequently, Faulkner's recommendations made little impact on the average farmer. There were some thoughtful agriculturists, among them Louis Bromfield, the noted author and conservation farmer in Ohio, who did believe that Ed Faulkner was correct, but that he was up against a centuries-old custom or practice that would be hard to dislodge. For that reason, Faulkner's plowless-farming idea was ignored by most farmers, and the plow continued to churn up the nation's soil.

Improved varieties of crops now permitted farmers to produce higher yields on less acreage. The development of hybrid corn, begun in the 1930's, was a major advance in crop cultivation.

Huge quantities of corn were needed to feed livestock as well as for other purposes. Corn Belt farmers, at first skeptical about hybrid corn, were now producing enormous yields by planting hybrid seed. In 1943, about 78 percent of all corn produced in the Corn Belt was of the hybrid variety.

Food demands for civilians and military personnel, at home and abroad, put great pressures on farmers. Grain producers, vegetable farmers, and livestock raisers doubled their efforts and operated at peak production. Still, there were shortages of certain foods. Meat, butter, eggs, and some vegetables were scarce commodities for most civilians during the war, since the armed forces were given priority on food. Food rationing was put into practice to prevent hoarding and to ensure a wider distribution of scarce foods.

No country in the world ever produced as much food and fiber—and in a relatively short time—as did the United States during World War Two. The fantastic rate of production was carried out with a reduced labor force. Farmers and farm workers were classified as essential to the war effort and as such were generally deferred from the draft. But many farm youths did go into the armed forces, and many farm men and women went to work in war industries. A labor shortage resulted, and skilled farm workers were in great demand.

To compensate for the lack of skilled farm workers, farmers depended more and more on mechanization. They also concentrated on producing more crops on fewer acres. But there was a reduction in the manufacture of powered farm machines to allow the military services more trucks, tractors, jeeps, bulldozers, and other heavy automotive machines. Steel and other critical metals were rationed, thus limiting the manufacture of farm equipment and implements. Experimental work on tractors was curtailed during the war because of the need to focus on the war industry.

Many farmers pooled their machines and equipment, with neighboring farmers sharing the use of tractors, combines, and other equipment, so that there would be fewer slowdowns in crop cultivation and harvesting. The shortage of farm machines and equipment gave rise to contract farm-machine operators, men who traveled from farm to farm with needed machines and equipment for hire. A few inventions were introduced during the war. They included the one-man hay baler, the cotton picker, and a hydraulic remote-control device on implements pulled by tractors.

When World War Two ended, American agriculture was a giant industry operating at top capacity. Again, as in the period after the First World War, farmers and government leaders expected America to continue exporting food and fiber for a long time. They believed that the terrible destruction to the lands, crops, livestock, and people wrought by World War Two would make a lot of foreign countries dependent on the United States for many years. Unfortunately, they miscalculated again. For a few years after the war, most European countries did rely on the importation of American food and fiber. But then a number of countries began to make a miraculous recovery as they rebuilt their impaired economies.

Various assistance programs sponsored by the United States were largely responsible for the rapid recoveries of war-torn countries. One of these was the restoration of agricultural industries destroyed by war. United States Military Government teams included veterinarians and crop and livestock specialists. Seeds and livestock were given to liberated or occupied countries and, with these as a foundation, Europe started to reconstruct its agriculture.

Gradually, the European countries recovered and began to depend less and less on America. And, once more, American farmers found themselves with surplus farm products. Prices fell off, but the costs of farm supplies continued to rise. It was quite evident that the farm boom days of 1940–1945 were gone and that farmers had to do something about balancing the scales of production and consumption of farm products.

In 1948, a Democratic President, Harry S. Truman, and a Republican-controlled Congress tried to shore up a sagging agricultural economy. Congress produced a weak version of the old Agricultural Adjustment Act, with low price supports to meet the problem of falling prices. No provision was made for funds to build more granaries and storehouses for the steadily mounting crop surpluses. Even though the war was over and the country was in the throes of changing back to a peacetime economy, farmers were still producing food and fiber at the accelerated wartime rate.

Foreign demands for American farm products kept tapering off, yet farmers had not yet cut back on their production. During the war, they had removed their poorest lands from cultivation and concentrated their farming on the best acres. Crop yields were also increased by better farming practices and more mechanization. Their problem now was to shift down to a lower gear of production to reduce surpluses.

Farmers in the decade after World War Two were far from being in the desperate condition they had endured in the 1930's. Yet many of them were uneasy about the future. The 1948 election returned Harry Truman to the presidency and brought in a Democrat-controlled Congress, which was more sympathetic to the needs of farmers. This Congress passed another Agricultural Adjustment Act, with high mandatory price supports. The government storehouses were filled with crop surpluses, and these cost the government heavy carrying charges. Manipulators and wheeler-dealers made high profits by exploiting the granary and storehouse situation. These profiteers became one of the scandals of the Truman Administration.

None of the emergency measures enacted during the Truman Administration offered any solution to the repetitive problem of overproduction and low prices. All that seemed to be accomplished was an increase in crop surpluses and government subsidies to farmers. Criticism was heaped on the farmers, and they were unjustly accused of being greedy exploiters who wanted government handouts at the expense of the taxpayers.

While the economics of agriculture were regressing, agricultural science and technology were moving ahead. With the removal of wartime restrictions on metals and other strategic materials, more farm machinery and equipment rolled from the factories. Some new machines and equipment were added to the impressive array of agricultural mechanization. Among the most important were a self-propelled corn picker and equipment to apply fertilizer in gas form (anhydrous ammonia).

Tractor production spurted upward, and in 1951 nearly 800,000 tractors came off the assembly line. The power takeoff, a device on tractors which permitted the powering of other machines pulled by tractors, became a standard piece of equipment. The International Harvester Company, manufacturers of farm machinery, came out with the Electrall in 1954. This was a farm electric power source, consisting basically of an electric generator mounted on a tractor and driven by the tractor engine. It supplied power for hay balers and for other uses for which electric power lines were not available.

Agricultural engineers were designing more safety features in tractors, which in the past had not always been safe. Some farmers had been severely injured or killed when tractors tipped over on steep slopes or were yanked over backward by heavy loads. Various professional engineering societies contributed to the development of more powerful, safer, all-purpose tractors in the 1960's. They included the American Society of Agricultural Engineers, the Society of Automotive Engineers, and the Farm Equipment Institute.

Another important development in farm mechanization was the invention and manufacture of more self-propelled machines. Engineers now produced combines, corn pickers, sprayers, hay balers, windrowers, and forage harvesters that operated under their own power. These new machines had several advantages, but the main one was that they released tractors for other farm work. This was important to those farmers who owned just one or two tractors.

New fertilizers, pesticides, herbicides, and other agricultural chemicals had become valuable tools in the farmer's continuous war against weeds and insects. A whole new industry had sprung up from the development of agricultural chemicals. DDT, a chlorinated hydrocarbon compound, which had proved so effective in insect control during World War Two, was now the world's leading pesticide because of its relative cheapness.

The character of agriculture in the United States was undergoing a transformation. It was moving away from the combination of family business and a way of life toward the status of big business. New and larger farm machines called for larger farm units and more efficient management along the lines of business organizations. The mammoth farm machines had to be used economically; they were not for the small or marginal farmer, for they were too expensive to operate on small acreages.

Also, the farmer with small acreage was turning out to be a liability in the 1960's. His farming methods were often inefficient and his costs of production were high. Except for part-time farmers with another source of income, few small-acreage farmers could compete with the large commercial farmers or the corporation-owned farms that were appearing throughout the country. The bulk of farm production for domestic and foreign consumption was being handled by about 20 percent of the nation's farmers, who farmed large units more efficiently and with up-to-date techniques.

A major result of the farm mergers and rise of corporation farms was a trend that took more people away from the farms. Thousands of farm workers gave up farming for other occupations. The number of farms had dropped from 6,814,000 in 1935 (the peak) to 4,857,000 in 1957. This exodus from the farms, although it reduced the labor force, did not drastically affect the total agricultural production. Farmers were producing more crops on fewer acres. However, the significant feature about the movement of people away from the farms lay in the fact that it emphasized the gradual demise of a famous and important part of American life: the family farm.

Agricultural education and technology continued to make advances in the 1960's. Colleges were turning out more agricultural graduates, and vocational agriculture was being taught in nearly ten thousand high schools. The agricultural colleges, experiment stations, and the Department of Agriculture conducted thousands of research projects aimed at increasing production on smaller acreages and giving the consumer a high-quality farm product. Crop cultivation, new and improved machinery, high-powered fertilizers, a variety of agricultural chemicals, superior livestock, top-yielding seeds, advanced fruit-culture practices, soil- and water-conservation programs, extensive irrigation projects—all of these areas and phases of agriculture were receiving the attention of scientists and technologists.

One hundred years after the establishment of the Department of Agriculture, farming in America had reached the highest level of production in the history of man. The agricultural revolution, which began in the nineteenth century, was entering a new stage. First had come the substitution of animal power for human labor. Then, in the early part of the twentieth century, mechanical power replaced most of the animal energy on farms. Now, in the 1960's, the third phase of the revolution—science and technology—was not only making farming more efficient and productive, but was improving the quality of life for millions of people all over the world.

Yet there were some flaws in the phenomenal growth of American agriculture. Farmers in the 1960's—like those in the late nineteenth and early twentieth centuries—were still victims of overproduction and low prices. Although they were independent by nature and occupation, they could not divorce themselves from government price supports and other subsidies. The problem of how to dispose of farm surpluses at fair prices still plagued farmers and government leaders, even though the best economic brains in the country had been grappling with the problem for many years.

Another serious weakness—or actually a side effect of agricultural growth—was that farmers were contributing to environmental pollution by mismanagement of soils, indiscriminate and wide use of agricultural chemicals, and the improper disposal of animal wastes. It was patent that the farmer was no longer an isolated part of conservation, an individual who had special problems that required special treatment. He was an integral part of the total environment, and what he did or did not do on his farm or ranch affected the rest of the world.

And as the 1960's ended it was also clear that the American

people as a whole were woefully ignorant of the values and needs of their great agriculture. There was an educational void—a severe communications gap that had to be eliminated if agriculture was to be accorded its rightful place in the American way of life. The public needed to be told that farm policy, a field subjected to constant opposition and criticism in the past, was formulated for the benefit of every citizen. Factors that affected farmers also affected the rest of the nation. And many of the agricultural research projects, roundly opposed by cut-the-spending groups, were of direct benefit to the consumers.

Americans also needed to be told that many farmers were disadvantaged, that they lived in relatively isolated areas and lagged far behind their urban cousins in opportunities for obtaining a higher standard of living. Farmers needed better schools, libraries, medical centers, and other necessities, or conveniences, of modern life. City people had to be informed that the per capita income of farmers had never caught up with that of the city workers. One quarter of all the farm families in America still had incomes of less than a thousand dollars a year. The tenant and sharecropper farms were still the rural ghettos of America. Finally, there was the need to recognize that the vanishing family farm was still the heart of American agriculture; that America's most valuable crop was her farm youth, who—because of farm depression—were abandoning agriculture as a business and a way of life.

11.

Farming Today

"Agindustry" is the new name of the game today. Modern farming is a huge, complex industry that affects millions of people on and off the farms. A gigantic food chain with many links, it begins with the farm supplier, who is followed by the main link, the farmer. Then the chain continues with the shipper, the processor, and the distributor, and finally ends with the consumer.

The character of American farming, which began to change after World War Two, has taken on the aspect of big business. Small family farms are becoming obsolescent and are giving way to large automated and mechanized agricultural production units.

Farms are not only larger today, but farmers practice more specialized production, for modern farming is a highly technical and competitive business, and no one person can be expected to be expert in all its branches. Once the average family farm produced a variety of crops and kept several kinds of livestock. Today, most farmers limit their production to perhaps no more than two or three crops or one or two kinds of livestock. Some farmers specialize even more, growing only one crop or even one variety of grain, fruit, or vegetable. Livestock husbandmen restrict their stock to one particular breed of steer, dairy cow, hog, sheep, or poultry. And most modern farm businesses are conducted on the principle of producing a greater volume on less acreage.

Large commercial farms are operated and managed by families or corporations along the lines of industrial plants. One farm worker produces enough food for himself and forty-three other people, a statistic which alone shows the remarkable advances made in agricultural production. However, this rate of production is achieved by only 20 percent of the nation's progressive and efficient farmers. There still are marginal and inefficient farmers who produce only enough food and fiber for their own or perhaps some local needs.

The commercial farmer is less self-sufficient than the farmer of thirty years ago. He produces none or at the most only a few of his needs on the farm, purchasing the bulk of supplies and services. He is a manipulator of technology and science, blending supplies, machinery, and scientific data to produce raw food and fiber.

Farming today requires considerable capital, and the risks are greater than ever. Operation costs are high; more than seventy cents of every dollar received by the farmer is spent on supplies and services. Modern farmers operate on a much narrower margin of profit than did their fathers. They do not make as much profit on a bushel of wheat, but the fact that they can produce a *greater volume on less acreage* means a larger net income for them.

Technology has done more than change the size of farms and cause farmers to specialize. Automation and mechanization have caused a surplus of more than three million farm workers. In addition to displaced farm workers, thousands are *underemployed.* On the large commercial farms, big machines—such as combines, threshers, corn pickers, forage harvesters, and vegetable pickers—have reduced the number of field hands necessary to harvest crops. In the grain-producing regions, new machines have cut the harvesting season in half. Farm laborers are hired only for short seasons or irregular periods. The reduced harvest seasons have forced many workers to move from farm to farm, and even from region to region, to obtain enough money on which to live.

The fluid condition of farm labor also affects the large farmers, for they have trouble in stabilizing and keeping a regular work force. Added to the instability of the migratory labor force is the fact that modern farming demands skilled workers. An able farm worker who has the knowledge and skill for efficient operation in one or more branches of farming is much sought after.

Specialization in farming has resulted in larger and more integrated production. Many of the commercial farms maintain management staffs similar to those in industry. For example, a corporation farm may employ a cost accountant, production manager, labor superintendent, foreman, salesmen, repair and maintenance men, and skilled farm workers. Some large farms even have chemists, engineers, plant specialists, and livestock nutritionists.

Automation and mechanization have truly revolutionized farming in America. They have not only increased production, but have eliminated the drudgery from many tasks and chores. Farmers also have more time for relaxation and recreation because of automation and mechanization.

Poultry production is one branch of agriculture that has a high degree of automation and mechanization. The average poultry farm of today bears little resemblance to that of twenty years ago. Assembly-line techniques have turned poultry farms into chicken and egg factories. Laying hens are kept in batteries of cages stacked on top of each other. Gone are the old poultry rooms or pens, where hens walked around most of the time, stopping only to lay eggs or to roost for the night. They are kept in confinement and on the job. Food and water are conveyed to them by automated equipment. The poultryhouse environment is controlled by automation. Eggs roll from the cages into troughs and are transported by conveyor belts. Droppings are carried away by special equipment.

Automation has contributed to the increase in egg production. The national average egg production is about 200 eggs per hen per year. Some large commercial flocks average 250 eggs per hen. More than 58 percent of the total poultry income is derived from the sale of eggs.

Poultry farmers who specialize in raising fryer and broiler chickens also profit from automation. So do turkey farmers. The elimination or reduction of time-consuming and laborious chores has speeded up the production of fryers, broilers, and turkeys.

There are, of course, other factors that have contributed to

A modern egg farm. *USDA Photo*

increased poultry production. Improved strains of chickens and better feeds are the keystones of progress in poultry farming. New hybrid strains of poultry, developed by scientists, yield higher quality eggs and meat. Balanced and fortified poultry feeds produce more eggs and meat sooner than did those of ten or fifteen years ago. Today, approximately 4.5 pounds of feed are required to produce a dozen eggs; ten years ago, 7 pounds of feed were needed to produce a dozen eggs. Better poultry feeds now make it possible to raise fryers and broilers to market weight (about 3.5 pounds) in nine weeks plus a few days. A decade ago, thirteen weeks were required to raise a bird to market weight. Also, fryers and broilers are being raised to market weight with less feed. Turkeys are also grown to market weight with less feed; approximately 3.5 pounds of feed per turkey pound are needed to raise a turkey to twenty-four weeks of age or the average market weight.

Functionally designed poultry houses, particularly those with ample lighting and controlled temperatures, also contribute to higher production of eggs and meat. Proper housing conditions are vital in egg production, especially when as many as fifty thousand hens are kept in one housing unit.

Automation and mechanization continue to increase efficiency on dairy and livestock farms. Livestock production per man hour has climbed at the rate of 3.7 percent a year during the last decade. Since dairy farming and poultry production have been traditionally the largest users of labor, the highest gains in labor

Beef cattle on a feeding lot in Nebraska. This lot has facilities to feed about six thousand head of cattle. *USDA Photo*

saved by automation and mechanization have been in these two branches of agriculture.

Beef consumption has almost doubled in recent years, a factor that has caused some changes in beef-cattle production. Superior livestock, improved feeds and feeding practices, automation and mechanization, and better transportation have all contributed toward giving the consumer quality beef.

Some new practices for feeding beef cattle have been introduced over the past few years. Steers have long been fed on corn; therefore, practicality and economy dictated that cattle be fed where the corn grows—in the Corn Belt. But now agricultural scientists have made it possible to fatten steers in regions other than the Corn Belt.

Milo, a grain plant, is grown on the Great Plains, and a new beef-cattle feeding center is developing there. Some meat packinghouses have already shifted their operations from the Corn Belt to the new center. First distributed to farmers in Texas, Oklahoma, Kansas, and Nebraska, milo has proved to be an excellent feed for beef cattle. As a result of planting milo, Kansas now has six times as many cattle on feed than it had ten years ago. Texas feeds more than 3 million head of steers, as opposed to about 200,000 animals ten years ago. Texas is second only to Iowa in the total number of steers fed for market.

Dairy farming, in terms of the value of its product, is the largest single agricultural industry in the United States. It, too, has been revolutionized by automation and mechanization. Better milking-machine systems, food conveyors, advances in artificial insemination, more efficient production of dairy-cattle feed and forage through the use of machinery such as hay balers and forage harvesters—these practices have all transformed dairy farming into an immense and highly productive arm of agriculture.

Hog production and sheep farming have to a lesser degree benefited from technological progress. However, all branches of the livestock industry have profited by scientific research. Animal sanitation has been improved by more potent germicides and detergents, as well as by more thorough cleaning of livestock equipment and facilities. A number of animal diseases have been eliminated; the incidence of others has been sharply reduced and kept under control by advances in veterinary medicine.

New methods of crop culture, although still in the experimental stage, promise easier and quicker production for grain and vegetable farmers. One such method, the no-tillage system for growing corn, is revolutionary. It consists of planting corn in unprepared soil; that is, seed corn is drilled directly into sod or stubble that has been killed by spraying it with a herbicide which is harmless to the corn plants. No-tillage farming eliminates plowing, disking, and harrowing. It calls to mind the recommendations of Edward Faulkner—doing away with the moldboard plow in preparing soils for cultivation. His idea met with indifference and criticism twenty-eight years ago. Now it has become a farm practice. The no-tillage system is being used in some Middle Atlantic States and in the eastern part of the Corn Belt. It shows great promise for preventing soil erosion on hilly lands, even though it seems to defy hundreds of years of corn culture.

In general, most crops have benefited from agricultural technology. In the last decade, corn yields have more than doubled, with an increase from thirty-eight bushels to seventy-nine bushels per acre. Cotton yields have increased; the average yield is now 511 pounds per acre, as opposed to 269 pounds ten years ago. Wheat yields have risen from 16.5 bushels an acre to 28.4 bushels, and soybeans have gone from 22 bushels an acre to 27 bushels.

The spectacular growth of American agriculture has created many jobs off the farms. Farmers are heavy users of manufactured goods. They spend more than $34 billion each year for goods and services necessary for the production of crops and livestock. Another $15 million is spent for food, clothing, and other consumer products. Farmers purchase various products that contain nearly 5 million pounds of steel, and use more than 320 million pounds of rubber on their machinery. They also use more petroleum products than any other industry, and require tremendous quantities of electricity.

This photo of a large farm near Los Angeles shows a contour-planted avocado grove on a steep slope (left foreground), citrus groves (center foreground), and pastures and grain land (background). The prevention of excess water runoff is one of the major problems in this area. *USDA Photo*

Notwithstanding his high rate of production, the modern farmer—like his father and grandfather—still has to contend with the old fears: drought, floods, insects, diseases, and soil erosion. The familiar problem of high costs of production and low prices hovers in the background. Farm incomes fluctuate from year to year, and even within a farming season. And a new problem has emerged to haunt the farmer: environmental pollution.

Farmers do not have absolute control over the quantity of crops or livestock that will be produced in a given season or period of years. Agricultural statisticians can say that the average wheat yield today is 28.4 bushels per acre, but there is no guarantee that a wheat farmer will obtain that yield. There are too many variables, too many factors that still make farming a gamble.

As a means of reducing the risks, a number of farmers work in conjunction with farm-related businesses, such as feed dealers, processors, or fertilizer companies. These nonfarming businesses share the financial responsibilities, as well as managerial decisions. This type of farming arrangement is known as *coordinated farming*. In a way, it is a kind of sharecropping.

Coordinated farming is used a great deal in commercial fruit and nut production. Also, about two thirds of the vegetables produced today are on a coordinated or contract basis, with the farmer assigning his crops to the large canneries or processors. Even poultry raisers have cooperative agreements with processors who supply most of the management. Poultry production now relies heavily on labor-saving or factory assembly-line techniques, which reduce the risks for both farmers and processors.

Another kind of farming that uses the contract method is sugar-beet production. Almost all the sugar-beet farmers enter into contracts with processors to ensure a market for their produce. Sugar beets are perishable and require prompt and careful handling from farm to processor. Farmers grow them under contracts that link the price of beets with that of sugar; the contracts usually specify how many acres of beets are to be cultivated, the kinds of seeds and culture, dates when the beets are to be delivered to the processor, and what marketing practices will be used. Some contracts also stipulate when the farmer will be paid for his beets. The chief value of this kind of farming lies in the fact that it lessens the risks for the farmer and more or less indicates what he may expect in the way of income.

Farm marketing practices have also undergone changes. This is especially true of those products consumed in the home. Most retail food stores are big supermarkets that stock over seven thousand items. Some of these supermarkets purchase farm products directly from farmers; others buy through wholesalers. Those farmers who sell directly to supermarkets receive better prices than those who must sell through middlemen or processors. A large number of farmers, however, sell their products through farm cooperatives. They also obtain their supplies from these cooperatives.

The cooperatives protect the farmers from unfair prices and

high costs of supplies. They build markets for farm products, thus making it easier for farmers to sell their produce. Cooperatives manufacture and sell farm supplies at reasonable costs to members, and they help farmers to obtain credit, electric power, insurance, irrigation, and other needed services. There are different kinds of cooperatives. Some specialize in farm supplies and insurance, some in marketing, and others in rural electrification. They range in size from small community cooperatives to large regional organizations, such as Agway Inc., which caters to the needs of farmers in twelve northeastern states.

Large regional cooperatives are organized along the lines of big business. Agway is a farm-supply and food marketing cooperative which is owned and operated by farmers in Connecticut, Delaware, New Jersey, Pennsylvania, Maryland, New York, Ohio, Massachusetts, Rhode Island, New Hampshire, Vermont, and Maine. Only farmers may own stock in Agway, and they must use Agway products and services.

This huge cooperative is a business corporation and pays federal income tax like any other corporation. Its primary object is to help farmers farm more efficiently and profitably. Agway produces and distributes farm supplies, such as fertilizers and feeds, and markets some of the products raised by its members. The cooperative maintains a thousand distribution outlets (stores, dealers, petroleum bulk plants, etc.), seventy production plants in the Northeast, and thirty processing plants and storage facilities. It conducts various research and development programs and gives grants-in-aid to agricultural colleges for research expected to be of value to Agway members. Although membership is limited to farmers, the general public is welcome, of course, to shop at Agway stores.

Inflation, rising labor costs, and increased taxes have made the cooperatives the farmer's best friends. Farmers—as individuals and even as a group—have little bargaining power or influence. Jonathan Davis, president of Agway, recently stated, ". . . as farmers, we no longer have the sheer numbers or perhaps even the sympathy of our state and federal lawmakers and government executives." And it is true: the mighty farm bloc has all but disappeared from the legislative scene.

Farm cooperatives provide competition that helps to lower the costs of farm supplies. They also secure for farmers the advantages of big business; for example, volume savings, managerial skills, capital, and the latest information about farming through educational programs and publications. Some cooperatives exert worldwide influence. For instance, the Sunkist Growers, Inc. and the Producers Export Company serve farmers in the farm export markets.

But the cooperatives, coordinated farming, and contract farming have not been able to eliminate the economic sore spots of agriculture. A decline in prices and an increase in costs of production still bring to the average farmer a net income that is less than that of the nonfarm worker. The per capita net income is about a thousand dollars a year; that of the average nonfarm worker is slightly more than two thousand dollars annually.

Farmers continue to be at an economic disadvantage for three main reasons: (1) the inability to adjust production to current demands; (2) farm costs climb faster than farm prices; and (3) there is still wide underemployment on farms because of automation and mechanization.

Many persons unfamiliar with farming and farmers often ask the question: Why don't farmers simply reduce their production when prices are low? A short answer is that farmers cannot cut back on their production without suffering a loss. Also, a cutback in production does not guarantee a rise in prices.

A number of farms produce materials that go back into the business; for example, most corn farmers raise corn to feed livestock. If they were to make drastic cuts in their livestock enterprise, then they would be reducing an outlet for their corn. It is important to bear in mind that farmers are dealing with living things, not automobiles, television sets, or furniture. A farmer cannot order a reduction in plant growth. True, he can plow under a crop or kill it with herbicides, but only at a great loss. Seed, fertilizer, labor, and time go into the cultivation of crops; these would all be lost if a farmer reduced production halfway through the season because prices were dropping.

Similarly, time and effort go into raising livestock. Sheep grow wool regardless of whether wool prices go up or down. Dairy cows produce milk irrespective of the price of milk. And so it goes. Farmers do not cut back production during periods of low prices because they cannot afford to do so without a great loss. Manufacturers do cut or reduce production because they cannot afford to do otherwise.

The search for parity continues, but the goal seems as elusive as ever. Farm prices today do have more stability than before the advent of price-support programs. Yet some agricultural economists believe that the goal should be parity of the farmer's *income*, rather than parity of farm product prices. Not all farmers favor price supports. The American Farm Bureau Federation, with 1.8 million member families, thinks that most price-support programs should be phased out of existence. It has recommended a program that would move farmers speedily to a free-market system by eliminating acreage limitations, allotments, marketing quotas, and compensation payments to farmers.

Yet other organizations and individuals think that price-support programs are still necessary. A sharp decline in wholesale farm prices—2.5 percent at the end of August, 1970—showed that the old problem of high production costs and low prices still remains. The possibility of increased labor costs and high transportation rates is another threat to farm incomes.

Along with its recommendations on farm price-support programs, the Farm Bureau advocates a law to establish ground rules for farm labor unions, including bans on product boycotts and harvest-time strikes. The call for a ban on boycotts and strikes is a result of more than three years of labor disputes and strikes by farm workers. One major strike and boycott problem is the movement by members of the United Farm Workers led by Cesar Chavez. This union, which originated among the grape workers in California's Central Valley, sought to win recognition and fair labor contracts from grape growers and other farmers.

One tactic used by the union was a boycott of grapes produced by growers who refused to go along with the workers' demands. Chavez managed to generate a great deal of sympathy for the grape pickers by leading a national boycott of nonunion grapes. After three years of the boycott and harvest strikes, including a fast by Chavez as a gesture of protest, the grape growers capitulated and signed contracts with the union.

Chavez's union and protest actions focused public attention on another farm labor problem: the migrant farm workers who travel from state to state seeking farm work. Even though the number of migrant farm workers has dropped from 400,000 to 250,000 in the last decade, the use of migrant workers probably will continue to be a significant part of the farm labor force for some time. These workers live in subhuman conditions in squalid migrant camps in New Jersey, Florida, and Texas. Farmers hire them because they are cheaper than machines.

Some on-the-spot observers have likened the lot of the migrant farm worker to that of the slave on southern plantations before the Civil War. Others have qualified this by stating that perhaps the slaves were better off, since their masters took better care of them than the present employers of migrant farm workers. The slaves were year-round property; migrant farm workers are on a farm only for a few months out of the year and the farmer does not want to spend money on bettering their living conditions.

Some attempts at bettering the conditions of the migrant farm workers have been made. New Jersey, for example, is the first state to try to eliminate some of the shocking conditions under which the workers live. Farmers in parts of Florida and Texas have attempted to do the same thing. But weak or nonexistent enforcement of sanitation laws, as well as of the civil rights of the migrant farm workers, still continues in many sections where this kind of labor is used.

In the spring of 1970, Iowa potato farmers went on strike for better prices. They burned more than twelve milllion pounds of potatoes in protest of low prices of potatoes. Labor strikes and potato bakes, along with deplorable conditions for migrant farm workers, serve to remind us that not all farmers and farm laborers are enjoying the riches of an affluent society—and that American agriculture, great though it may be, still has some serious weaknesses.

On the brighter side of agriculture, farm exports have risen to more than five billion dollars in the 1965–1970 period. Several factors account for this rise in exports, including improved economic conditions in some industrial countries and unfavorable crop-growing conditions in others. The United States now supplies about one fifth of the total volume of farm products in world trade. Farm exports are grown on one acre out of every four in production in the United States. American farmers export wheat, feed grains, cotton, cottonseed, tobacco, rice, fruit, vegetables, and animal products.

More than 125 countries import American farm products, but the largest importers are Japan, Canada, the United Kingdom, India, Italy, Spain, West Germany, the Netherlands, Poland, and Belgium. These countries purchase about two thirds of the total American exports. The United States Government also finances a Food for Peace program, which provides funds to a number of developing countries. One third of our surplus crops is given to needy countries. The other two thirds are sold, and foreign currency from the sales is given to the governments as a grant. These grants can be used for family planning or national defense (as in South Vietnam).

Science and technology have made American agriculture the envy of the world, yet there are still some rough spots beneath the veneer of prosperity and achievement. Stabilized farm prices, lower costs of production, better wages for farm laborers, improved living conditions for migrant workers, the retraining of workers displaced by automation and mechanization, and protection of the environment—these are challenges for farmers today and tomorrow.

12.

Science and Agriculture

Agriculture, although regarded as a science, actually is a composite of three scientific disciplines: the *life, physical,* and *social* sciences. Many scientists from these three disciplines are engaged in research and development geared to finding new ways to increase farm production, obtaining maximum use of food and fiber, conserving natural resources, protecting the environment, and raising the standards of living for millions of people. Agronomists, botanists, chemists, ecologists, economists, engineers, entomologists, foresters, meterologists, physicians, sociologists, veterinarians, and zoologists are some of the scientists who are making notable contributions to agriculture.

The fountainhead of agricultural research is the Agricultural Research Service of the United States Department of Agriculture. Home base for nearly five thousand scientists is the Beltsville Research Center in Maryland, located about fifteen miles northeast of Washington, D. C. Established in 1910 on 475 acres purchased by the Department of Agriculture, the Beltsville Center now oversees more than three thousand research projects conducted there and at other centers and laboratories.

Beltsville's acreage has been increased from the original 475 to the 10,500 acres which now make up the experimental fields, pastures, gardens, orchards, and woods. More than a thousand buildings house offices, laboratories, livestock, poultry, machine shops, a heating plant, a library, and an auditorium. Thirty-five greenhouses spread out over five acres. Over three thousand head of livestock, including cattle, swine, sheep, and goats are main-

Plant geneticists at Beltsville, Maryland, are developing new varieties of potatoes. *USDA Photo*

tained on this huge research farm. Chickens and turkeys number more than ten thousand, and the Center keeps about three thousand small animals, such as hamsters, guinea pigs, and rabbits for research purposes.

In addition to the main research center at Beltsville, there are regional research divisions which conduct basic and applied research into new and wider uses of farm products. These regional divisions were created in 1938 and are known as Utilization Research and Development Divisions. They are the modern counterpart of George Washington Carver's farm-product research laboratory at Tuskegee Institute. The utilization research divisions are composed of laboratory complexes that engage in research of their own and in cooperation with agricultural experiment stations at the land-grant colleges and industries.

The Eastern Utilization Research and Development Division is located in Wyndmoor, Pennsylvania, a few miles northwest of Philadelphia. Scientists at this division do research on animal products such as milk, cheese, fats, and hides, and on plant products including eastern fruits, tobacco, and maple sap.

A Northern Utilization Research Division is situated in Peoria, Illinois. Here, research is centered on corn, wheat, barley, grain sorghum, oats, oilseeds, soybeans, flaxseeds, and new crops. There is a National Animal Disease Laboratory at Ames, Iowa, where veterinarians and other scientists concentrate on the control and eradication of infectious diseases of livestock.

The Southern Utilization Research Division has its headquarters in New Orleans, Louisiana. Research at this center focuses on cotton, cottonseed, pine gum, southern fruits, vegetables, sweet potatoes, peanuts, rice, and cucumbers.

Western fruits, nuts, vegetables, poultry products, forage crops, wheat, rice, barley, wool, mohair, dry beans, peas, and castor and safflower oilseeds are the subjects of research at the Western Utilization Research and Development Division laboratories at Albany, California, near San Francisco.

In addition to the major utilization research and development centers, the Department of Agriculture operates various field stations and laboratories scattered around the country.

Everyone concerned with agriculture—farmer, rancher, transporter, processor, packager, distributor, wholesaler, retailer, and, of course, the consumer—benefits from the work and the contributions of the Agricultural Research Service. The research is programmed toward providing more quality food and fiber despite diminishing acres, finding new ways of producing abundant food at more economical costs, and developing new and better fibers. It is true that America has an abundance of food and fiber today. But what about tomorrow with its increased population? Agricultural research is one way to ensure that there will be sufficient food in the future.

Agricultural science is more or less divided into three major fields of research and development: *production* of food and fiber, *utilization* of farm products, and *processing* and *marketing* of food and fiber.

Since farmers must produce on relatively fewer acres today—and do so despite continuing attacks on crops and livestock by insects and diseases—the efforts of agricultural scientists are directed toward making the work of the farmer easier and efficient. All kinds of research projects are being carried out in laboratories and field stations of the Agricultural Research S ice. These are supported and supplemented by work in experiment stations at agricultural colleges. In a way, research into farm production attempts to eliminate, or at least ease, some of the farmer's old fears: crop failure, soil erosion and infertility, insects, animal and plant diseases, and floods.

Soil fertility and protection are major areas of agricultural research. High crop yields can be produced only on fertile and non-eroded soils. Agricultural scientists constantly improve the quality and potency of farm fertilizers, develop new ones, and conduct experiments on the best methods of application. For example, during the past few years, midwestern farmers have learned to grow more wheat with the use of sulfur and nitrogen fertilizers.

Investigations have shown that livestock production can be increased in the northern Great Plains by applications of fertilizers to grasslands. Previously, fertilization of grasses was not very successful, mainly because there was no way to prevent loss of valuable nitrogen. Erratic or insufficient rainfall in the Great Plains prevented the fertilizers from being properly used by the grasses. Now scientists have devised new application techniques that allow the grasses to reap the benefits of nitrogen and other essential elements in fertilizers.

Scientists have also discovered new ways to make use of livestock wastes. As most farmers know, farm wastes, particularly manure, are excellent fertilizers. Before the development of commercial fertilizers, animal manures were the chief source of fertilizers. Manure is still used for this purpose, but some farms are producing more than they can use. This is especially true of those farms on the fringes of cities and suburbs, farms that do not have enough croplands or pastures on which to spread the manure. Consequently, manure piles up, and heavy rains or spring thaws carry this farm waste into streams, ponds, lakes, and rivers. Thus, farm wastes contribute to environmental pollution.

USDA plant specialists have developed new varieties of grains. From left to right: wheat, rice, and barley. *USDA Photo*

Agricultural Research Service engineers have designed special lagoons to trap and store liquid farm wastes, and plant specialists have grown forage grasses in these lagoons, employing a type of hydroponic farming. The grasses grown in the lagoons flourish and return part of the waste nitrogen (which is present in the manure runoff) to livestock by way of protein in the grasses.

Another important area of research on farm production is that of increasing the hardiness and productivity of crop and forage plants. Scientists at the Tifton, Georgia, Agricultural Research Service laboratories have developed a new Bermuda grass hybrid that has the potential of increasing the daily gains of cattle fed on it by as much as 30 percent. These scientists have also produced a high-yield forage hybrid—a cross between pearl millet and napier grass—that remains succulent and abundant well into November, the time when most forage plants start to dry up and drop in yield.

Several new crops that show a greater resistance to diseases and insects, as well as having higher yields, have been developed by agricultural scientists. They include three new varieties of wheat, a hybrid barley, two new varieties of cotton and rice, a new kind of oats, soybean, Korean lespedeza, and a blue lupine. One of the new varieties of wheat—an improved type of Gaines wheat—shows great promise of high yields to add to the world's food supplies.

Henbar, the new variety of barley, was developed in Arizona. It has yielded up to 35 percent more grain per acre than the leading variety previously grown in Arizona. There is another unusual feature to Henbar barley: it is the first commercial hybrid barley ever developed. The barley plant's genetic makeup has prevented hybridization in the past. But through countless breeding experiments, plant geneticists have finally mastered a technique for producing hybrid barley. This technique may lead the way to the development of hybrids in vegetables and field crops that have so far resisted hybridization by the usual methods.

Some valuable breeding projects are also under way in the livestock industry. Sheep normally lamb once a year. Now scientists have developed a new strain of sheep—the Morlam—which lambs every eight months. Morlam ewes have produced an average of two lambs per pregnancy, or six lambs every two years. This is a significant increase in lamb production.

Methods of increasing beef yields are another area of scientific investigation by agricultural scientists. It has been shown that systematic crossbreeding, rather than straight linebreeding, results in quicker beef yields. Crossbreeding experiments produced 3 percent more calves, which were heavier at weaning time than linebred calves. Scientists believe that crossbred calves are more vigorous than purebreds; also, crossbred calves grow faster because of hybrid vigor resulting from new combinations of genes.

Geneticists are now experimenting on *interspecies* breeding; that is, the breeding of animals of different species. A promising interspecies cross is one between a Dark Cornish chicken and a Japanese quail. Although of different species, both the chicken and the quail are members of the order Galliformes. The offspring of this cross is intermediate in size between a chicken and a quail. Scientists believe that continued crossing and selecting may eventually produce a bird that tastes like a quail but is the size of a chicken.

Size and taste, however, are not the most important aspects of this interesting cross. What is important is the combining of genes that never before have been combined by nature. Interspecies breeeding opens up a whole new line of possibilities in the field of livestock breeding.

If farmers are to increase and maintain high crop yields and healthy livestock production, they must have the means to control insects and diseases. Some of the most damaging insect pests are the boll weevil, the grasshopper, and the alfalfa weevil. These three insects alone cause the loss of millions of dollars in crops destroyed. The alfalfa weevil is so destructive in some regions that the crop can no longer be grown profitably. Alfalfa is the chief forage crop; it provides the raw materials that cows convert into milk. A loss in alfalfa yield means a corresponding loss in milk for the dairy farmer.

A new breed of sheep—the Morlam (more lambs)—produces lambs in shorter time than other breeds. The ewe seen here gave birth to these lambs, two sets of twins and a single, between May, 1965, and September, 1966 (only eight months between births). *USDA Photo*

There is a constant war between agricultural scientists and insect pests. One tactic developed by scientists against the alfalfa weevil is to burn the insects out of alfalfa fields by using flames on the plants when they are dormant, which is in the late fall and winter. Field demonstrations have shown that this flame method will kill 90 percent of weevil larvae in a field of alfalfa, 80 percent of weevil eggs, and 55 percent of adult weevils.

Plant breeders are working to develop varieties of alfalfa that will be highly resistant to attacks from weevils. This program is a long and arduous one. It requires the screening of thousands of alfalfa plants to find a few that are weevil-resistant. From these few plants, scientists hope to develop a weevil-resistant variety.

Insect control is a top-priority program on the Agricultural Research Service agenda. The focus is on finding ways to control insects without the use of chemicals that are harmful to human beings and wildlife. Long-range environmental pollution by pesticides now makes it imperative that farmers use safer and less toxic insect-control methods.

The Agricultural Research Service's insect-control program is known as TIPS (total insect population supression). In the past, the majority of insect-control programs were directed toward the reduction of insect numbers to a point where they no longer inflicted great or extensive damage to crops and livestock. When the damage that could still be done by any remaining insects *was less than the cost of totally eradicating them,* then the control program was considered successful. Today, the TIPS program has a larger objective: total elimination of certain insect populations in a community, state, region, or even throughout the United States.

There are approximately ten thousand species of insects in the United States that may be called enemies of agriculture. One hundred of these cause extensive damage to crops, more than 80 percent of all insect damage. Twelve species—including the boll and alfalfa weevil—are responsible for half the total damage to crops by insects.

Various insect-control methods are under investigation by TIPS scientists. One is the use of black-light traps. Insects are lured into the traps by the black light. This technique was tested on the tobacco budworm on St. Croix, Virgin Islands. Despite its name, the tobacco budworm attacks other plants, including cotton and tomatoes. Fortunately for us, the budworm moths are easily attracted by black light and will fly toward it into any traps. More experiments are being conducted on the use of black-light traps, and if this method of insect control proves feasible on a large scale, it will be recommended to farmers.

The most promising and revolutionary insect-control measure, however, is the release of sterile insects into areas containing wild, fertile insects of the same species. Its great value lies in the fact that it reduces or eliminates insect populations without the use of pesticides. The originator of this technique is Dr. E. F. Knipling, director of entomology in the Agricultural Research Service. This technique has already eliminated harmful insects in certain regions.

One insect that has been eradicated is the screwworm fly. Until the development of Dr. Knipling's method, the screwworm fly was a highly destructive insect pest that preyed on livestock and wild mammals in the southern United States. The fly laid its eggs in cuts and other wounds of cattle, sheep, deer, raccoons, bears, and, on occasion, human beings. Maggots hatched from the eggs and attacked the flesh of the hosts, practically eating them alive. Screwworm flies were so destructive to deer in Florida and Texas that more than a third of the annual fawn crop succumbed to these voracious pests.

Dr. Knipling and his co-workers raised millions of screwworm flies in special fly factories in the laboratory. The flies were rendered sterile by the use of radioactive cobalt. Then clouds of them were air-dropped over infested regions. Sterile males outnumbered the native males and mated with the wild females. Eggs from these matings failed to hatch. More sterile males were dropped, and the numbers of screwworm flies gradually declined until the pests were totally eradicated in the region.

This biological insect-control technique has a potential for not only eliminating agricultural insect pests, but those attacking wildlife, as well. Deer populations in Florida have risen more than 60 percent since the screwworm-fly eradication program was started. Similarly, the survival rate of fawns in Texas has doubled as a result of the eradication program.

Other nonchemical methods of insect control are under investigation by TIPS scientists. Natural parasites that attack the alfalfa weevil and other insects are being used to suppress these

A grasshopper's potential for damage is illustrated in this close-up of a nymph (note undeveloped wings on back) feeding on corn. This specimen belongs to the two-striped grasshopper species. *USDA Photo*

pests. In the case of the alfalfa weevil, the parasites are small wasps that lay their eggs in weevil larvae. When the eggs hatch, the young wasps kill the weevil larve and eat them.

Grasshoppers are ubiquitous insects. There is hardly a region or community that does not have hoppers—insects that have plagued man's crops ever since plant cultivation started some nine thousand years ago. The grasshoppers attack all kinds of crops, including wheat, corn, barley, rye, oats, flax, alfalfa, and many forage crops.

So far, the main method for controlling grasshoppers has been by the use of insecticides. However, TIPS scientists are searching for biological controls for the hoppers. Researchers in one Agricultural Research Service laboratory found that a disease organism, *Melamoeba locustae,* did kill grasshoppers exposed to it in the laboratory. Of course more research is necessary, but this disease organism may provide an important control for the pestiferous hoppers.

The infamous boll weevil of the South causes millions of

dollars of damage to cotton each year. An immigrant from Mexico, the boll weevil quickly and firmly established itself in the southern United States, from California eastward through the humid sections of the Deep South. Boll-weevil control programs have been based mainly on the use of pesticides, but the spraying of cotton fields is costly and has not always been successful.

Agricultural scientists have studied the habits of the boll weevil and the effect of various pesticides on the insect. They learned that the boll weevil readies itself for winter by entering a resting stage, known as the *diapause.* In this stage the boll weevil is immune to cold and other adverse conditions. Since the boll weevil can be killed more easily *before* the insect enters the diapause stage, boll-weevil spraying programs are more effective when this fact is taken into consideration. The scientists have developed methods for spraying smaller amounts of potent pesticides, which reduces the costs of treating a cotton field.

Sterilization as an insect-control measure is also being used to eradicate boll weevils. In the spring of 1970, TIPS scientists

Close-up of a boll weevil on a cotton plant. Boll weevils destroy millions of dollars' worth of cotton annually. *USDA Photo*

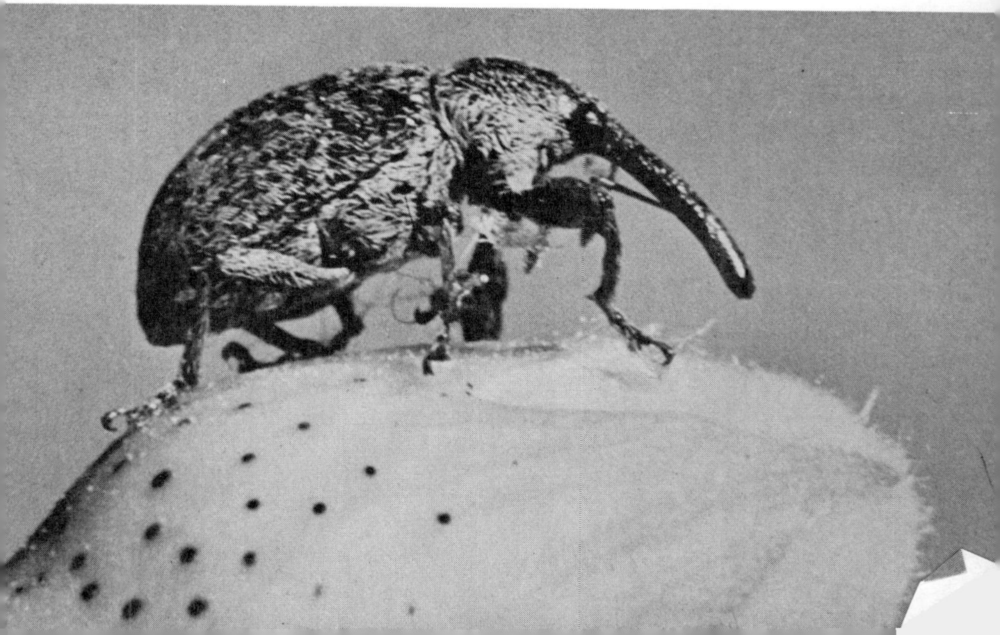

released 192 million sterile bollworm moths over southern California cotton fields. The sterile insects (male and female) offset the reproductive cycles of the natural or wild bollworm moths. And although bollworm moths mate several times during a season, the invasion by millions of sterilized moths greatly reduces the chances of fertilization. If a wild female bollworm moth is mated first with a sterile male and then with a wild fertile male, the results of the second mating are still greatly reduced insofar as producing fertile eggs are concerned.

The Mediterranean fruit fly was a multimillion-dollar threat to fruit and vegetable production in the Rio Grande Valley. Now, because of the sterile-insect release program, the region is free from the fruit fly. Infestations of the khapra beetle and citrus blackfly, both foreign pests, have been checked by the same technique.

Animal disease control is another major field in which Agricultural Research Service scientists are at work. Nine states are now free from hog cholera, thanks to the research of scientists and the cooperation of livestock farmers. Thirteen states have eliminated brucellosis, a bacterial disease of cattle, sheep, swine, and goats; this disease is also transmissible from animals to human beings. Sheep scabies, once a major problem, is all but eliminated. Many turkey flocks are now certified as being free from pullorum disease, which is highly fatal for young birds and formerly caused large losses. Two other turkey diseases—paratyphoid and infectious sinusitis—are targets for veterinary and other agricultural scientists.

Many other research programs and projects, too numerous to list here, are in progress in the laboratories and field stations of the Agricultural Research Service. This agricultural research is not one-sided or conducted solely for the benefit of farmers. Other branches of the giant agindustry are being improved by research and development. The consumer is, of course, the ultimate beneficiary of all the research.

Scientists at the Utilization Research and Development Di-

Equipment for popping or puffing fruit or vegetable pieces. The pieces are partly dried and then shot from a gun (shown at top). Explosion puffing creates tiny channels in the pieces, allowing them to soak up water before cooking.

vision laboratories keep searching for new "faces" for farm products. They have come up with many, and the average American is familiar with most of them: low-fat cheese, boneless turkey roasts, instant potatoes, frozen fruit juices, low-cost broiler and fryer chickens, dry beans that are ready to eat in thirty minutes, peanuts high in protein but low in fat, dried apple flakes, freeze-dried strawberries, popped cereals—these are some of the new farm products developed by agricultural scientists.

Utilization research is contributing toward clean streams in America. The development of detergents from animal fats that are completely biodegradable and don't cause foaming is one way agricultural scientists are helping to protect the environment. Another is finding new and profitable uses for whey, a by-product of cheese manufacture that can no longer be discarded as waste because of its potential as a pollutant. Still a third way is the recovery and utilization of substances in the wastes from the processing of agricultural products; scientists have discovered techniques that will both lower the polluting effect of the waste and provide processors with more useful products to sell.

Agricultural Research Service scientists have developed two superior wash-and-wear finishes now widely used in the textile industry. Stretch cotton is a product of agricultural research. On the market today are shirts, slacks, shorts, and other clothing items that will not bag, but will stretch or give with strain or pressure. Scientists have also put stretch in cotton lace. Other improvements have been made on cotton, such as resistance to water, mildew, and stains; minimum shrinkage and fading; longer-lasting pleats and creases in wash-and-wear clothes; brighter, more lustrous colors; and improved cotton batting for auto seats and mattresses.

A lesser known but valuable contribution of agricultural science is the mass production of antibiotics. Microbiologists in the Agricultural Research Service developed a quantity-production technique for penicillin during World War Two, when great quantities of this life-saving antibiotic were needed. A researcher discovered a more productive strain of penicillin, which became the parent stock for all modern cultures. A mass-production technique that involved the use of fermented corn wastes and milk sugar to foster the growth of the penicillin mold was discovered, and this led to the large-scale production of penicillin and other antibiotics.

Agricultural scientists continue to work on a variety of projects that may save human lives. A leech repellent developed by scientists protects United States soldiers from land and water leeches in Asian jungles. Experiments are being conducted on Spanish moss as a possible cancer-inhibiting material. A substance found in a Chinese tree may someday provide the means of arresting leukemia. Poultry pathologists were the first to point out that a virus can cause cancer. Research on poultry cancers may yield vital data that can be applied to human cancers.

The union between science and agriculture has proved to be a powerful force in strengthening the links in the human food chain. Thousands of research and action programs, embracing every aspect of modern agriculture and some nonagricultural areas, improve the production, utilization, and quality of food and fiber. Agricultural science deals not only with better food and fiber today, but is working to ensure an abundance of quality food and fiber tomorrow.

13.

The Farmer and the Environment

A hundred and fifty years ago, most Americans believed that the country's natural resources were unlimited. America was a big, sprawling land with vast forests, countless soil types, huge deposits of minerals, many waterways, and an abundance of wildlife. The land was regarded as a bottomless treasure chest of natural resources that were there to be developed and used to meet the needs of a growing nation.

With this credo as a guide, America's natural resources were not only used but exploited with a wanton recklessness and waste. Forests were leveled with ax and saw to clear land for farms, towns, and mills. Prairie lands were churned by plows and sliced by disks and harrows to grow crops. Marshes and wetlands were drained and converted into more croplands. Wildlife was killed with traps and guns, and those birds, mammals, and amphibians that survived were deprived of necessary habitats as agriculture and industry expanded. The decline of the wilderness paralleled the growth of America, for the natural resources of the continent were the raw materials from which a mighty nation was built.

By its actions and lack of controls, the federal government aided and abetted the wasteful exploitation of natural resources. It failed to provide adequate controls and safeguards, and it fostered the destruction of the environment with various exploitation incentives. Settlers rushed to plow up the western lands or to stock the ranges with cattle and sheep under the Homestead Act. Railroads were handed choice lands that were the habitats of many species of wildlife and the ancestral hunting grounds of

A trumpeter swan in Red Rock Lakes Refuge, Montana. Swans and other waterfowl nest and feed on farm ponds and other waterways in rural regions. *U.S. Department of the Interior, Fish and Wildlife Service*

Indians. Lumber companies were allowed to waste timber and loot large tracts of public forests with cut-and-get-out practices, with no thought to reforestation. Cattle and sheep ranchers were permitted to overstock the grasslands and pave the way for soil erosion. Swamps were drained or filled in under the provisions of the Swampland Act of 1849, thereby displacing more wildlife and otherwise disturbing the delicate ecology of submarginal lands. Cotton and other single-crop farmers were encouraged to produce bumper harvests without any programs to replenish the fertility of overworked soils. Few if any people thought about protecting the nation's waters. The making of America took much from the land, and only a handful of men thought about putting something back.

In 1864, George Perkins Marsh, a Vermont lawyer and philologist, warned about the dangers of tampering with the land and upsetting the balance of nature. In his book, *Man and Nature or Physical Geography as Modified by Human Action,* Marsh stressed the need for a new approach in dealing with nature—the approach which we know today as ecology. He pointed out that to disturb the balance of nature without considering the consequences was to invite disaster. Marsh compared life to a huge web that was composed of every single organism. Any drastic change in, or destruction of, this web of life could disrupt the entire biological community. All life was interdependent, and the loss of one species affected the welfare of others.

Marsh and his ecology were ahead of the times, and only a few thoughtful men realized what he was writing about. Toward the end of the nineteenth century, some voices were raised in protest over the continued destruction of the wilderness and the waste of natural resources. Through the influence and efforts of eminent men such as Theodore Roosevelt, Gifford Pinchot, William Temple Hornaday, and John Muir, a conservation movement was born. It began slowly and with difficulty, for there was strong opposition from vested interests and apathy on the part of the general public.

Gradually, the conservation movement gained momentum and strength. More and more Americans became aware that the wilderness was not endless, that their wildlife and other natural resources were not limitless, and that wasteful practices and exploitation were slowly but surely robbing them of a most valuable natural heritage. Some regulatory steps were taken to halt the exploitation of natural resources and to preserve some of the wilderness, but these were mostly in the area of wildlife and forest protection. Hardly any attention was given to the preservation of the nation's farmlands.

Few people even thought of farms as a necessary part of the conservation movement. But there were some persons who saw the folly of unwise farming practices that depleted soils and rendered them susceptible to erosion. Secretary of Agriculture James Wilson was one nineteenth-century agriculturist who warned about the wasteful soil practices. George Washington Carver was another. However, even these men did not visualize farms and farmers as being integral and indispensable parts of the total environment. The preservation and improvement of soils were, in the opinion of most agriculturists of the period, measures taken to increase agricultural production. That soil and water should be protected and used wisely as a means of protecting the total environment did not occur to early agricultural conservationists.

Twentieth-century Americans inherited the results of mismanaged farmlands and wasted natural resources. They were handed stark evidence that natural resources were not limitless. A sharp decline in timber yields, an alarming drop in the populations of wildlife and the extinction of some species, soils impoverished by intensive farming without fertilization, severely eroded soils in the South and Southwest, western grazing lands destroyed by overstocking with cattle and sheep, destruction of natural sponges by draining swamps and marshes, waste of water resources—all of these scars on the land attested to the fact that farmers, the federal government, state governments, and other

An example of water pollution in a stream.

stewards of the land had failed to adequately protect the environment and natural resources.

The New Deal agricultural and conservation programs were the first concerted government efforts to protect and manage natural resources. But even these did not restore the balance of nature or reverse the destructive trend. To a certain extent, they did slow down the exploitation and made some repairs to the damaged continent. And they made more Americans aware that the nation's farms were part of the total environment—vital strands in what George Perkins Marsh had called the web of life.

Thirty years later, in the 1960's, most of the old conservation problems remained, and new ones had appeared. Air and water pollution, contamination of the environment by pesticides, chemicals, and radioactive fallout, pollution from animal wastes, and a steady loss of wilderness and wildlife—all were side effects of technology and urbanization.

These new environmental problems, along with the old, forced a revision of the over-all concept of conservation. The original ideas of preservation and protection of natural resources were still valid, and they were reaffirmed. Resource management—in addition to the soil conservation begun in the 1930's—was also recognized as a major conservation tool. But it was now clear that there had to be another approach to conservation, one that would protect the total environment of air, water, soil, plants, wildlife, livestock, man, cities, suburbs, wilderness, and farms. The emphasis of corrective action had to be on environmental protection and improvement, not just on the conservation of natural resources. Thus, one hundred years after it had been proposed by George Perkins Marsh, the ecological approach to dealing with nature was activated.

Today, environmental protection and improvement are the keystones of conservation. Most enlightened people now realize that soil erosion, air and water pollution, indiscriminate use of pesticides, loss of wildlife, and urban blight affect *everybody*, regardless of where they live or what their station is in life.

This view of a farm in Mills County, Iowa, shows a complete terrace system. In the left background is a waterway that has been recently worked in.
USDA Soil Conservation Photo

In the last decade, a great deal of progress has been made in conserving natural resources in rural America. Soil-conservation programs, initiated by the Soil Conservation Service in the 1930's and implemented by measures of the Soil Bank in the 1960's, have improved the fertility of many soils and checked erosion in many regions. Ninety-six percent of all farms and ranches (which comprise more than 1.6 billion acres) are in locally governed soil-conservation districts. More than 700 million acres have been mapped by the Soil Conservation Service, and over 70,000 soil types have been examined and analyzed for character and condition. Ninety-nine million acres of farmland are in soil-conservation systems that favor protective measures such as strip farming, contour farming, and terracing.

Many farmers have strengthened soil structure and fertility

by disking or plowing under crop residues on more than 68 million acres. Thirty-eight million acres of farmland are in protected woodlots, which serve as important components of watershed conservation programs. Water runoff from crop fields and orchards, especially in hilly regions, has been reduced on a million acres by the use of grass waterways—natural watercourses or strips of land in crop fields that are seeded with grass to reduce the flow of water from heavy rains and to halt erosion. There are over a million miles of terraces and more than 25 million acres of drainage improvements. Six million acres of hilly or steep farmlands have been leveled for more efficient land use.

More efficient farming practices have also helped to protect the land. Millions of acres in the Southeast have been taken out of cotton and other soil-depleting row crops and converted to pastures for dairy and beef cattle. The modern progressive southern farmer revitalizes his soils instead of working them to death. He uses more lime and fertilizer than did the farmer of thirty years ago. In the West, controlled grazing systems reduce the threat of grass shortages and soil erosion. Formerly eroded grazing lands have been planted in suitable and nutritious soil-retaining grasses.

Valuable prairie lands are being preserved by conservation-minded individuals and organizations. Several midwestern universities are reclaiming stretches of prairie land for scientific, social, and aesthetic purposes. A national conservation agency, the Nature Conservancy, is buying bits of prairie land in Illinois, Kansas, Nebraska, Minnesota, and Wisconsin. In Kenosha, Wisconsin, some citizens purchased about eighty acres of prairie land to keep it from becoming a victim of the land developers' bulldozers. And many conservationists, farmers among them, are urging the federal government to establish a Prairie National Park as a further means of preserving the dwindling grasslands.

Although there have been victories in the battle to protect and restore the land, the war is far from being over. Soil erosion is still a major problem. Nearly four billion tons of silt are deposited each year in streams, rivers, ponds, and lakes—soil that is washed off farmlands and highway embankments. Soil erosion is not just a rural problem; poorly designed and hastily built housing developments and supermarket parking lots contribute to soil erosion. In building these facilities in cities and suburbs, land developers cut down trees that help to control water runoff, and they remove valuable topsoil, thus exposing subsoil or hardpan, which is carried away by heavy rains and floods.

Unfortunately, too many farmers persist in using tillage practices that lead to soil erosion. They still plow and cultivate up and down hilly fields, instead of following the natural contour of the land. Some tobacco farmers in Maryland use this harmful method of tillage. Also, many farmers fail to use strip farming and grass waterways to protect their soils.

Years ago, most farm fields were laid out in straight lines. The main reason for this was that it made surveying easier. Farmers just naturally plowed and cultivated along straight lines. But up-and-down hill plowing along straight lines leads to erosion. When heavy rains fall, the bare ground in between row crops, such as corn and potatoes, is loosened by the tremendous accumulative power of the raindrops—millions of them pounding down and bouncing up with soil particles. The raindrops, now tinged with soil, develop into puddles, which in turn become rivulets that surge downhill, gaining speed and carrying soil, manure, fertilizers, and pesticides across roads and into creeks, rivers, and other waters. The accumulation of soil and farm wastes—often laden with disease organisms—flows through cities and into esturaries and finally into the sea.

Soil and water are vital to life as well as to agricultural production. Properly protected and managed, they support life in our communities and on our farms, but neglected or unrestrained, they weaken or break many food chains, including that of man. Of these two essential natural resources, water is considered to be the basis of any land-conservation program. When water is efficiently managed, the other natural resources dependent upon it—soil, plants, livestock, and wildlife—are protected. This concept is the foundation of land management today; it provides the guidelines for the conservation of farm resources on a watershed basis. Each natural area of land that drains into a creek, pond, lake, or river is treated and managed as a unit in the over-all watershed program.

This practical, comprehensive land-management approach is the basis of the National Watershed Protection and Flood Prevention Act. Farmers and other landowners are encouraged to follow conservation practices which create absorption areas or giant sponges that soak up as much water as possible *at the site where it falls,* whether it is a cropfield or orchard. Surplus water —that which runs off the land—is trapped in upland reservoirs *before* it can cause serious erosion that damages soil, plants, and animals. Watershed protection is a most important environmental protection measure and one in which more farmers and landowners should participate.

The wide and indiscriminate use of pesticides, herbicides, and other agricultural chemicals continues to bring forth loud protests from national and local conservation agencies. There is too much evidence of the damaging effects to the environment by pesticides and other chemicals for farmers to shrug off the protests as coming from nature cranks. Pesticides, particularly the long-lasting chlorinated hydrocarbon compounds, have drastically disturbed the ecology of many plant and animal communities. In the case of DDT, an almost omnipresent pesticide, many organisms in the web of life have become contaminated, including man, Practically every person who has lived through the period from 1950 to the present time has at least a trace of DDT in his or her system. Because of pressure from conservation groups and the realization that DDT has outlived its usefulness (many insects have developed a resistance to this chlorinated hydrocarbon pesticide), the federal government has banned its use, except in certain controlled situations.

Herbicides—plant- or weed-killing chemicals—have helped to reduce labor in crop cultivation, to increase yields, and to eliminate unwanted plants. But herbicides, like pesticides, pose a threat to wildlife and the public health. One herbicide, 2,4,5-T, has been condemned by a number of conservation agencies as being hazardous to the public health. Scientific studies have shown that 2,4,5-T has a potential for *teratogenesis*, or the ability to cause monstrosities or abnormal formations in plants and animals. The Biometrics Research Laboratories at Bethesda, Maryland, reported in 1968 that 2,4,5-T produced a number of birth defects when fed to mice and rats. In 1970, the USDA banned 2,4,5-T for use in home gardens and on food crops.

There are, of course, many more agricultural chemicals; some are more toxic than others, and some have a longer residual effect than others. However, even the least toxic agricultural chemicals must be used with a view toward their immediate and long-range effect on the public health and environment. Farmers should become familiar with the properties and hazards of all agricultural chemicals used on their farms and follow the manufacturer's directions explicitly.

If any area of modern agriculture presents a challenge, it is the problem of controlling insects and weeds without damaging the environment. Obviously, if farmers are to continue producing sufficient food and fiber, they must control insects and weeds. No intelligent person would contest this action. What can be contested, though, is environmental pollution by pesticides, herbicides, and animal wastes.

Some conservationists have accused farmers of not only polluting the environment by unwise and overuse of pesticides and other chemicals, but of being indifferent to the damage done. The accusation is unfair. True, some farmers may not care whether they pollute the environment, but most of them are genuinely concerned about the problem. However, they are caught in a trap or are victims of a paradox. They are expected to keep producing high yields of food and fiber, yet are criticized for using the pesticides and chemicals that have enabled them to achieve the highest farm-production rate in the world. To fight insects and weeds, farmers are using whatever means have been recommended to them by the federal government. They have demonstrated their willingness to learn. When new and safer methods of insect and weed control are made available, farmers will use them.

Agricultural scientists are seeking less toxic chemicals for use on farms. They are also studying the ability of plants to modify chemical substances. It is also important to know how much of a pesticide or herbicide ends up in the edible parts of food plants. Experiments at Fargo, North Dakota, have revealed that crops do modify pesticides; in some cases, plants have actually converted pesticides into compounds that are relatively harmless to livestock.

Although these results have been obtained only in the agricultural research laboratories, they may be duplicated on the farms in the near future. In the meantime, farmers must exercise care and caution in the use of agricultural chemicals if they wish to avoid more regulative and repressive laws or actions. If the levels of chemicals in fruits, grains, vegetables, milk, eggs, and meats reach harmful proportions, then the human food chain is endangered. When this happens, man's control of his food chain will have backfired on him, for he will have climaxed thousands of years of agricultural progress with the production of contaminated food.

Animal wastes are another main source of farm pollution. A large portion of animal waste comes from the dairy and beef industries, although other livestock farming areas are contributors to environmental pollution. Many beef-cattle feed lots and dairy barnyards slope toward streams, and heavy rains carry manure and other contaminators down the slope into the water. The amount of waste from a single steer is considerable. Cecil H. Wadleigh, director of the Soil and Water Conservation Research Division, Beltsville, Maryland, stated in a talk before the 1970 Wildlife Management Institute conference that an average thousand-pound steer "excretes 110 pounds of nitrogen, 22 pounds of phosphorus, 125 pounds of potassium, and 365 pounds of biochemical oxygen demand [substances that use up oxygen in water] in a year." He pointed out that on a two-hundred-head beef-cattle feed lot of one acre, there is an accumulated excretion of 11 tons of nitrogen, more than 2 tons of phosphorus, 12 tons of potassium, and nearly 37 tons of biochemical oxygen demand—all of which adds up to a highly polluted acre. When water runoff carries animal waste from this acre to the nearby streams, it adds to the buildup of environmental pollution.

This is just one situation in which animal waste contributes to environmental pollution. There are countless other instances throughout the country. The average dairy farmer spreads manure on his fields and pastures in the fall and winter, even on top of snow. During thaws and heavy spring rains, water runoff carries the animal wastes into streams and other waterways. Similar pollution originates on swine and poultry farms, especially when the manure or droppings are piled outside and left to the actions of rain and water runoff. For example, wastes from Long Island duck farms have been one of the major causes of pollution in Long Island Sound. As a resut of pollution from duck-farm wastes and other sources, the production of shellfish has been greatly reduced in the Sound.

Animal waste from unprotected manure piles finds its way into ponds, streams, and rivers, thus adding to the buildup of environmental pollution.

USDA Photo

Some corrective measures are being taken to prevent animal wastes from adding to environmental pollution. Oxidation lagoons —pools or reservoirs that trap and oxygenate animal wastes—are helping to cut down on pollution from animal wastes. But these lagoons must be operated efficiently and receive regular maintenance. The lagoons, like sewage treatment plants and septic-tank systems, can malfunction or become inoperative; when they do, pollution is back again.

Farmers can help reduce pollution from animal wastes by not allowing manure to accumulate on sites exposed to heavy rains or floods. Conservation farming practices and soil-erosion control will also help to check pollution from animal wastes. Some farmers dry and package excess manure for sale to plant nurseries and home gardeners, thus reducing the chances of environmental pollution.

One might ask: If farmers are contributing to environmental pollution with soil erosion, chemicals, and animal wastes, why isn't more being done to stop them?

First of all, farm pollution, like industrial pollution, is a complex problem. There are no quick and easy cures for either one. A great deal of agricultural pollution might be stopped by abandoning practices that contribute to pollution. This is both an extreme and ridiculous suggestion, although it has been made. It would mean eliminating or drastically reducing farm production in many areas, an action that would be tantamount to shutting down or cutting back the output of vital industries and services. Neither of these two measures is a realistic or practical solution to the problem of environmental pollution.

Then what can be done?

To be realistic, a farm-pollution clean-up program, like that of industrial pollution, probably has to be sponsored and supported by the government. Such a proposal, of course, invites criticism in that it would be just another government handout to farmers. But the truth of the matter is that farmers cannot do the job themselves; they are simply too poor.

The farmer's income is still far below that of the city worker.

A large amount of soil erosion and animal-waste pollution comes from the poorer farms. Farmers in the Ozarks, Appalachia, and other agriculturally disadvantaged regions have incomes that are below the national poverty level. They cannot even begin to think of pollution control. The idea of eliminating pollution on their farms to reduce environmental pollution downstream or miles away in cities and suburbs is a concept that many poor farmers do not understand. Their personal problems take precedence over other things. All of their labors and scarce funds must go into eking out a bare existence. And even the larger and more affluent farmers in the rest of the country lack the means to combat farm pollution on their own.

Environmental education and technical-assistance programs are needed at the grass-roots level. Farmers, of course, should join in the battle against environmental pollution in whatever way they can. Farm labor unions, cooperatives, and national farm organizations should police their own industry. These things should be done not only to silence critics but because farmers have a responsibility to the land and the environment and to future generations.

Many farmers are already at work in the environmental clean-up and conservation of natural resources. Some of the soil-conservation programs and watershed projects have been mentioned. More than half of America's wildlife habitats are located on farms and ranches. Some farmers are helping to protect and manage wildlife by creating and improving habitats. Crop planting for wildlife food, maintenance of hedgerows and other shrubs for cover and nests, construction of farm ponds for waterfowl and aquatic mammals as well as for recreational purposes, efficient management of woodlots for wildlife cover and watershed protection, and rearing of game birds for later release in the wild—these are some of the ways by which farmers *are* contributing to a better environment.

Farmers are stewards of the land. Most of them are deeply concerned about protecting the environment, and they realize that more must be done to preserve nature. But farmers have a realistic attitude regarding the immensity of the job, unlike some of the more militant conservationists and environmental protectionists. Farmers know that if they must do the job alone, their costs of production will rise still further. They also know that farm-pollution control may cause a slowdown in the current trend toward higher crop and livestock yields because of the curtailment of certain practices. And farmers think that if they must clean up farm pollution by themselves, the costs of doing so, like those of industrial-pollution control, may ultimately be borne by the consumers.

Farms and farmers are indispensable parts of the total environment. They, too, would welcome a rural America enhanced by conservation farming, abundant wildlife, clean streams, and better housing and living conditions for farm families. And they are willing to be shown the way to achieve such an environment.

14.

Agriculture in the World of Tomorrow

Will Americans have enough food in the years ahead—for instance, by the year 2000?

Of course nobody can predict what is going to happen ten, twenty, or thirty years from now, certainly not with any certainty or accuracy. But by projecting present food production and consumption trends to the near future, some kind of pattern may be obtained. But even this pattern does not guarantee any long-term prospects for a plentiful food supply thirty years from now. Agriculture, as mentioned before, is—and will be—subject to variable factors. Weather, insects, diseases, soil erosion, droughts, and floods—any one or a combination of these can drastically affect food production.

As proof of the risky nature of farming we have but to recall the damage done to the corn crop of 1970 by a fungus blight. Despite all of the great improvements in corn culture—hybridization, weather resistance, and other advantages—an estimated 5 percent of the nation's corn crop was attacked and damaged by a fungus called *Helminthosporium maydis*, which rotted the leaves of the corn plants. This fungus was not a newcomer; it had been around for a long time. Still, most corn plants, particularly the hybrid varieties, managed to survive the attacks of this pest. Then, according to mycologists, the fungus mutated into a more virulent form and corn plants were unable to withstand attacks in the summer of 1970, which happened to be an unusually humid season in most of the corn-growing regions.

The corn blight had an immediate effect on the farm pro-

Farmlands are diminishing near large cities and suburbs as housing developments and industrial sites use up valuable acreage.

Photo by Joseph Morsello

duct market, sending all grain futures upward. Demand for grain futures (grain bought and sold for delivery at a future time) is based on a kind of domino theory that takes in livestock. For example, when corn becomes scarce as human and animal food, there is an increase in the use of oats. Also, farmers have to feed wheat to livestock and poultry in larger amounts to offset the loss in corn. More soybeans are converted into meal and this is used as an additive in livestock feeds. Consequently, less grain is available for direct human consumption.

There are some other important factors that will have a bearing on future food supplies. These include the rate of the population growth, whether the present and projected productivity of farms can be maintained, the degree of efficient use of animal and plant foods, development of supplementary sources of food, foreign demands for American farm products, and environmental pollution and waste of natural resources.

Certainly the population growth will directly affect production and availability. Although there has been a decline in the birthrate since 1955, the expectation (based on current trends) is that it will increase during the last part of the century. The United States Census Bureau estimates that the population will reach 400 million by the year 2000. Other statistical sources warn that the population may go as high as 700 million by that time.

An increased population will mean a corresponding increase in food consumption. In turn, farmers will have to increase their production. As a comparison with the expected population increase, the Department of Agriculture estimates that farm production will double by the end of the century. Also, the increase in farm production will be achieved with a *smaller* work force, the use of *more* land, and with a *greater* volume of nonfarm resources.

The Economic and Statistical Analysis Division of the United States Economic Research Service has set a figure of 1.5 million farms by the year 2000, if present consolidation trends continue. Economists suggest that less than half of these farms will be responsible for the major portion of commercial food production. An important factor to be considered in future food production is that millions of acres of croplands are now being kept out of production to hold down crop surpluses. If a food shortage should occur ten, twenty, or thirty years from now, these lands could be put back into the agricultural production line.

Agricultural research toward developing higher-yield crops, improving livestock, and finding new uses for food will, of course, help to ensure at least the potential for an ample food supply in the years to come. A better and wider utilization of farm products could include a more complete use of photosynthesis of plants, which—it will be recalled—is the basis of all food chains. Actually, only a small portion of the total yearly output of photosynthesis is now utilized for food.

The leaves of some plants, for example, which are now discarded could be used as food. Although some people eat beet and carrot tops, these parts of the plants are usually thrown away; yet they have nutritive values that can supply energy and protein. There is also the possibility of a more complete utilization of chlorophyllic sea algae as a source of food. Sea algae is supposed to be a hundred times more efficient than corn in using solar energy.

There has been a great deal of talk about future food needs being provided from the sea. But sea resources, like other natural resources, are not limitless and must be protected and wisely managed to ensure sufficient yields. Estuary and offshore pollution must be halted if there is to be a continued supply of seafood. In short, the sea can be a major source of future food supplies, provided that source is not destroyed today.

America's commercial fisheries are a complex industry, which has never been developed to its fullest potential as a supplementary source of food. The United States consumes about 11 percent of the world fish catch; 70 percent of all fish consumed by Americans is imported, with 60 percent of tuna, 50 percent of shrimp, and 90 percent of spiny lobsters coming from abroad.

There is no question that an expansion of our fisheries will provide an important adjunct to farm production. Some specific fishery areas that could be developed further are shrimp off the Maine coast, scallops near Alaska, and abalones in California waters, just to name a few. A unique kind of farming is being done in Louisiana, where farmers are raising crawfish or "crawdads" in flooded rice fields. The crawfish are relatives of the larger lobsters, and farmers are experimenting with them as a new rotation crop with rice.

Many of our inland lakes and rivers are polluted, but it should be possible to increase inland fisheries if water pollution is reduced or halted. Fish farming or the production of fish in artificial environments is another future source of food. Research teams in Scotland, for instance, have shown that fish farming is practical. One research team has been breeding and fattening fish in tanks constructed on land near nuclear power plants. The warm waste water from the plants is used for growing fish. Flatfish (fluke, flounder, sole, etc.) raised in these tanks have grown at *twice* the rate of those in the sea. These Scottish fish farmers have succeeded in producing marketable fish in about eighteen months. One reason why these tank-reared fish have such remarkable growth rates is that the warm water extends the active feeding season in the spring and autumn.

Fish farming might be a partial answer to waste water from nuclear and other power plants in the United States. Conservationists, marine biologists, and ecologists now warn that thermal pollution caused by warm water entering normally cool waterways will harm certain fish and aquatic plants.

Another factor affecting future food supplies is the degree of competition between domestic and foreign demands. These, in turn, will be governed by the rate of production, and even though the decline in the number of farms is expected to continue, the rate of production promises to remain high. But if a slowdown in production occurs, then the competition between domestic and foreign demands for American farm products will be keen.

All of this leads to a larger question: Can the world feed itself in the years ahead? It is a question that the leaders of the United States and other countries are asking their agricultural and technical advisers. Hunger is a world problem, one that requires international cooperation if it is to be eliminated in the world of tomorrow.

In the past fifteen years, some developing countries, largely through the help of United States aid programs, have managed to increase their food production. The introduction of science and technology, new high-yield varieties of wheat, rice, and other grains, larger applications of fertilizers, more irrigation, wider use of mechanization, better farming practices, and improved livestock—these are some of the advances that have enabled emerging countries to produce more food. But so far none of the developing countries has reached its peak in agriculture. Each one has a long way to go before being self-sustaining, and each relies on help from the United States and other developed countries.

Again, nobody can predict what will happen to the world

food supply in the years ahead. World agricultural production depends on much the same factors as does American farming. One thing is certain, though: The United States will be producing the bulk of the world's farm products.

American farmers have been exporting their products since colonial times. But our farm exports have usually been purchased by those industrial countries which could afford them. Only since the 1950's have American farm products been extended to disadvantaged or poor countries. Public Law 480, enacted in 1954, authorized the Food for Peace program, whereby needy countries were given credit with which to obtain American farm products.

Since the Food for Peace program began, American farmers have shipped more than 135 million metric tons of wheat and 40 million tons of other grains to needy countries. This program has allowed developing countries to increase their food consumption even though their food production has lagged behind their minimum needs.

America unquestionably will have to continue the Food for Peace or similar programs in the next decade. However, dependence on this kind of aid probably will diminish as developing countries improve their economy to the point where they can either produce enough food for their needs, or where they will be in a position to purchase extra food supplies on the open world market. To achieve this state of relative food independence, the developing countries must accelerate their agricultural production as well as improve their general economy. Some new countries are already doing this; others have a long way to go.

Finally, the condition and quality of tomorrow's total environment and natural resources will markedly affect the food supply. We have belatedly started to clean up environmental pollution and to stop the waste of natural resources. But the damage has been too extensive, too widespread for any quick repairs. Many years will be required to restore the environment to a healthy state, if indeed it can ever be restored.

Smog and other air pollution definitely and adversely affects crop production, particularly on farms close to cities and industrial centers. Airborne radioactive materials also pose problems today and unless they are checked will be a problem tomorrow. Radioactive materials contaminate the leaves and stems of plants. Although the levels of radioactive materials have rarely reached harmful stages, it will take constant vigilance to see that they never do reach injurious proportions.

Scientists are studying plants that are sensitive to air pollution; for example, tobacco, oats, and alfalfa. Also under investigation are *antioxidants,* chemical substances which, when sprayed on plants, prevent injury from certain air pollutants. Agricultural scientists have discovered ways to prevent soybeans and grain sorghums from taking up radioactive materials from the soil. They have also devised a method of washing radioactive materials from

A farm pond with contour strip-farming nearby (Harford County, Maryland).
USDA Soil Conservation Service

wheat, and they have managed to remove about 90 percent of radioactive materials from milk. These and other important antipollution measures will have to become standard procedures in tomorrow's agriculture.

Soil and water conservation must be a vital component of any future farming system. Soil erosion must be checked and fertility maintained; water must be clean, stored, and wisely used. The average rainfall over the United States is approximately thirty inches. Some regions receive more or less rain, of course, with some areas receiving as little as two inches a year. We are not likely to be blessed with more rainfall from the atmosphere, not unless some unusual meteorological phenomenon occurs. Fortunately, there is enough water to sustain life and, if managed wisely, enough left over for other uses. But water resources will have to be kept clean and free from harmful pollutants.

Undoubtedly, water management will play a mojor role in the farming system of tomorrow. It will be necessary to redistribute water to those regions with scant rainfall. Farmers will have to pay close attention to the establishment and maintenance of watersheds; they must devise methods for trapping and storing rainfall, instead of allowing it to run off with valuable topsoil; they need to practice more efficient irrigation and reclaim wetlands for crop production or wildlife habitats.

There is no reason to believe that there will be a food shortage by the year 2000. However, this does not mean that Americans should become complacent about food production. There is hunger in America today and hunger throughout the world. The biggest challenge to our young future farmers is to abolish world hunger. Young Americans who are now studying agriculture in the colleges and vocational schools will have the know-how to produce enough food in the years ahead. But it is up to the rest of American youth to see that food is properly distributed and that no one goes hungry.

Glossary

Achira: a tropical plant with edible roots.

Agave: any of the plants in the amaryllis family, including the common fleshy-leaf century plant and sisal; sisal leaves yield a strong white fiber used for making cordage.

Agindustry: the new and collective term for all phases of agriculture: farmers, suppliers, processors, distributors, etc.

Agronomist: a scientist who specializes in the production of crops and soil management.

Agrostology: the branch of botany that deals with grasses.

Biometrics: the science of statistics pertaining to animals and plants.

Botanist: a biologist who specializes in the study of plants.

Buckwheat: an herb of the genus *Fagopyrum* cultivated as a food crop.

Cacao bean: the fruit of a South American tree, *Theobroma cacao*, used chiefly in making cocoa and chocolate.

Combine: a machine which harvests, threshes, and cleans grain while moving over a field.

Contour farming: A system of plowing and cultivating a field in lines or arcs *across* a slope, rather than up and down; contour furrows hold or slow down water runoff.

Corn picker: a self-propelled machine that harvests corn.

Crossbreeding: a system of breeding involving the mating of two varieties of the same species, as in corn, livestock.

Ecology: the science that deals with the relationships of animals and plants to each other and their environment.

Einkorn: a one-grained wheat with a short, flat, barleylike spike.

Emmer: a wheat grown in Russia, Germany, and north-central North America, especially as a livestock food.

Forage harvester: a machine that cuts and chops forage plants, such as alfalfa, Sudan grass, etc.

Gypsum: hydrous calcium sulfate; used for top-dressing soils.

Hog cholera: an acute, highly infectious virus disease of swine, characterized by dullness, diarrhea, and temperatures to 107 degrees F.

Horticulture: the science of growing plants, especially vegetables, fruits, and ornamental plants.

Hydroponics: the art of growing plants, especially vegetables, with their roots immersed in water containing all of the essential plant nutrients; also known as "tray agriculture."

Lentil: a Eurasian plant of the pea family grown for its edible lens-shaped seeds.

Linebreeding: a system of breeding in which closely related animals or plants are mated; in livestock breeding, mating cousin to cousin would be an example of linebreeding.

Manioc: any of several plants of the spurge family grown in the tropics for their fleshy edible rootstocks which yield starch; also known as cassava.

Marl: an earthy, crumbly deposit of clay mixed with calcium carbonate and used as a fertilizer on soils that are deficient in lime.

Meteorologist: a scientist who specializes in the study of atmospheric conditions, such as heat, moisture, wind, etc.

Millet: any of various small-seeded cereal and forage grasses, such as pearl and Italian millets.

Mohair: the hair from Angora goats.

Mycologist: a scientist who specializes in the study of fungi.

Naval stores: raw materials and supplies used in shipping, such as turpentine, resin, tar, etc.

Neolithic: pertaining to the New Stone Age.

Oca: a tuberous plant (*Oxalis*) used for food.

Paleolithic: pertaining to the Old Stone Age.

Paratyphoid: an infectious disease of turkeys and other poultry that is caused by many strains of *Salmonella* bacteria. It is characterized by trembling, gasping, loss of egg production, dehydration, and diarrhea.

Pepino: Spanish for cucumber; also a yellow fruit grown in highlands of Peru.

Pleuropneumonia: a highly infectious disease of cattle, characterized by a high temperature (as high as 107 degrees F.), rapid and shallow breathing, cough, thirst, and loss of appetite.

Pullorum disease: an infectious bacterial disease of poultry, characterized by a whitish diarrhea and loss of appetite.

Rinderpest: an infectious disease of cattle, sheep, and goats, characterized by inflammation of the membranes, especially of the intestines.

Safflower: an Old World thistlelike herb of the aster family; the flowers from this plant are used for red dyes and drugs.

Scabies: a mangelike disease of sheep and other animals.

Sinusitis: an infectious disease of the lower part of the bony socket of the eyes. The symptoms include swelling of the head and a nasal discharge.

Sorghum: any of a genus of tropical cereal grasses cultivated for fodder and sugar.

Strip farming: a system of cultivation in which grasses and row crops are alternated in strips; the grasses are planted uphill from row crops, thus preventing water from running over exposed soil in the row-crop strip; may be used in combination with contour farming.

Tedding: the practice of turning new-mown grasses for drying.

Terracing: a system of steplike ridges of earth raised at right angles to a slope; soil is taken from both sides of the ridges, leaving adjacent ditches to catch and move aside runoff water.

Texas fever: a disease of cattle caused by a species of protozoan and transmitted by the tick *Boophilus annulatus;* characterized by a temperature rise, often as high as 107 degrees F., loss of appetite, and marked depression.

Thresher: a machine used to separate grain from the stalks.

Vetchling: any of a genus (*Lathyrus*) of small herbs.

Windrowers: a machine used to arrange hay or straw in rows prior to being picked up by a baler.

Suggested Reading

Agricultural Economics:

Mellor, John W., *The Economics of Agricultural Development*. Ithaca, New York, Cornell University Press, 1966.

Agricultural Literature:

Crèvecouer, Hector St. John de, *Letters from an American Farmer*. New York, E. P. Dutton, 1957 (paperback).

Agricultural Machinery and Power:

Power to Produce. U.S. Department of Agriculture Yearbook. Washington, D. C., Government Printing Office, 1960.

Animal Science:

Acker, Duane, *Animal Science and Industry*. Englewood Cliffs, New Jersey, Prentice-Hall, 1962.

Briggs, H. M., *Modern Breeds of Livestock*. New York, Macmillan, 1969.

Diggins, Ronald V., and Bundy, Clarence E., *Livestock and Poultry Production*. Englewood Cliffs, New Jersey, Prentice-Hall, 1968.

Biography:

Holt, Rackham, *George Washington Carver: An American Biography*. New York, Doubleday, 1963.

Kruif, Paul de, *Hunger Fighters*. New York, Harcourt, Brace, 1928.

Lowenthal, David, *George Perkins Marsh, Versatile Vermonter*. New York, Columbia University Press, 1958.

Careers in Agriculture:

Agriculture Is More Than Farming (booklet). Alexandria, Virginia, Future Farmers Supply Service.

Hoover, Norman K., *Handbook of Agricultural Occupation*. Danville, Illinois, Interstate Printers and Publishers, 1969.

Phipps, Lloyd J., *Your Opportunities in Vocational Agriculture*. Danville, Illinois, Interstate Printers and Publishers, 1962.

Dairy Farming:

Diggins, Ronald V., and Bundy, Clarence E., *Dairy Production*. Englewood Cliffs, New Jersey, Prentice-Hall, 1961.

Ecology:

Kormondy, E. J., *Concepts of Ecology*. Englewood Cliffs, New Jersey, Prentice-Hall, 1969.

Storer, John H., *The Web of Life*. New York, New American Library, 1966 (paperback).

Environment:

Dasmann, Raymond F., *Environmental Conservation*. New York, J. Wiley & Sons, 1968.

McCoy, J. J., *Shadows Over the Land*. New York, Seabury Press, 1970.

Farm Labor:

Bishop, C. E., *Farm Labor in the United States*. New York, Columbia University Press, 1967.

McWilliams, Carey, *Factories in the Field: The Story of the Migratory Farm Labor in California*. Hamden, Connecticut, Shoe String Press, 1969.

Matthiessen, Peter, *Sal Si Puedes—Escape If You Can: Cesar Chavez and the New American Revolution*. New York, Random House, 1969.

Farm Magazines:

American Agriculturist, *Better Farming Methods*, *Farm Home and News*, *Farm Journal*, *Farm Quarterly*, *Hoard's Dairyman*, and *Successful Farming*.

Fiction:

Mitchell, Margaret, *Gone With the Wind*. New York, Pocketbooks, 1967 (paperback).

Richter, Conrad, *The Fields*. New York, Bantam, 1965 (paperback).

Steinbeck, John, *The Grapes of Wrath*. New York, Viking, 1958 (paperback).

Young, Stark, *So Red the Rose*. New York, Charles Scribner's Sons, 1969 (paperback).

Food and Hunger:

Food for All of Us. U.S. Department of Agriculture Yearbook. Washington, D. C., Government Printing Office, 1969.

Kotz, Nick, *Let Them Eat Promises*. Englewood Cliffs, New Jersey, Prentice-Hall, 1969.

FRONTIER DAYS:

Hawgood, John A., *America's Western Frontiers*. New York, Alfred Knopf, 1967.

Turner, Frederick Jackson, *The Frontier in American History*. New York, Holt, Rinehart and Winston, 1963.

FRUIT FARMING:

Schneider, G. W., and Scarborough, C. C., *Fruit Growing*. Englewood Cliffs, New Jersey, Prentice-Hall, 1959.

GRAIN FARMING:

Delroit, Richard, and Ahlgren, Henry J., *Crop Production—Principles and Practice*. Englewood Cliffs, New Jersey, Prentice-Hall, 1953.

Wolfe, T. K., and Kipps, M. S., *Production of Field Crops*. New York, McGraw-Hill, 1953.

HUMAN POPULATION:

Ehrlich, P. R., *The Population Bomb*. New York, Ballantine, 1968 (paperback).

IMMIGRATION:

Kennedy, John Fitzgerald, A *Nation of Immigrants*. New York, Harper & Row, 1969 (paperback).

Handlin, Oscar, *The Uprooted*. New York, Grosset and Dunlap, 1957.

IRRIGATION:

Hargreaves, Mary Wilma, *Dry Farming in the Northern Great Plains*. Cambridge, Massachusetts, Harvard University Press, 1957.

MACHINERY:

Jones, Fred R., *Farm Gas Engines and Tractors*. New York, McGraw-Hill, 1963.

Power to Produce. U.S. Department of Agriculture Yearbook. Washington, D. C., Government Printing Office, 1960.

Smith, H. P., *Farm Machinery and Equipment*. New York, McGraw-Hill, 1948.

MAIZE

Mangelsdorf, Paul, "Ancestors of Corn." *Science*, Vol. 125 (1958).

Wissler, Clark, "Corn and Early American Civilizations." *Natural History*, Vol. 54 (1945), pp. 56–65.

MORMONS:

Carmer, Carl, *The Farm Boy and the Angel*. New York, Doubleday, 1970.

New Deal Days:

Chase, Stuart, *Rich Land, Poor Land*. New York, McGraw-Hill, 1936.
Schlesinger, Arthur M., *Age of Roosevelt*, Vol. II, *The Coming of the New Deal*. Boston, Houghton Mifflin, 1958.
Wallace, Henry Agard, *New Frontiers*. New York, Reynal & Hitchcock, 1934.

Pesticides:

Carson, Rachel, *Silent Spring*. Boston, Houghton Mifflin, 1969.
McMillen, Wheeler, *Bugs or People?* New York, Appleton-Century Crofts, 1965.

Rangelands:

Donahue, R. L., Evans, F. Everett, and Jones, L. J., *The Range and Pasture Book*. Englewood Cliffs, New Jersey, Prentice-Hall, 1956.
Harlan, Jack R., *Theory and Dynamics of Grassland Agriculture*. Princeton, New Jersey, Van Nostrand, 1956.

Rural Sociology:

Farmers in a Changing World. U.S. Department of Agriculture Yearbook. Washington, D.C., Government Printing Office, 1940.

Soil and Water Conservation:

Bennett, Hugh H., *Elements of Soil Conservation*. New York, McGraw Hill, 1955.
Faulkner, Edward H., *Plowman's Folly*. Norman, University of Oklahoma Press, 1963.
Sears, Paul, *Deserts on the March*. Norman, University of Oklahoma Press, 1967.
Soil. U.S. Department of Agriculture Yearbook. Washington, D.C., Government Printing Office, 1957.
Water. U.S. Department of Agriculture Yearbook. Washington, D.C., Government Printing Office, 1955.

Vegetable Farming:

MacGillivray, J. H., *Vegetable Production*. New York, McGraw Hill, 1952.
Thompson, H. C., and Kelly, W. C., *Vegetable Crops* New York, McGraw Hill, 1957.

Wildlife:

Allen, Durward L., *The Farmer and Wildlife*. Washington, D.C., Wildlife Management Institute,
McCoy, J. J., *Saving Our Wildlife*. New York, Crowell Collier, 1970.
Trefethen, James B., *Wildlife Management and Conservation*. Chicago, D. C. Heath, 1967 (paperback).

Index

(Figures in italics refer to illustrations.)